Resilient Computer System Design

Victor Castano • Igor Schagaev

Resilient Computer System Design

Springer

Victor Castano
IT-ACS Ltd
Stevenage, UK

Igor Schagaev
IT-ACS Ltd and London Metropolitan University
Stevenage, UK

ISBN 978-3-319-38605-8 ISBN 978-3-319-15069-7 (eBook)
DOI 10.1007/978-3-319-15069-7

Springer Cham Heidelberg New York Dordrecht London

Softcover reprint of the hardcover 1st edition 2015

Printed on acid-free paper

Springer International Publishing AG Switzerland is part of Springer Science+Business Media (www.springer.com)

Preface

New areas of ICT applications require complete redesign of computer systems to address challenges of extreme reliability, high performance and power efficiency. Up to now there are no consistent concepts, theories and texts, enabling us to design systems with mentioned requirements. Requirements themselves became processes and evolve along life cycle of the systems and applications. All these force us to start all over again, leaving past and sentimental values behind, making new computer systems and software that match mentioned requirements and new applications.

Our 35 years of experience consolidates: design of airborne computers and black boxes; parallel computers for submarines, helicopters and satellites. Our intensive research work in academies and higher education institutions included analysis of performance, reliability and design of computer systems. Both regretfully have proved at various levels that existing computer system and system software solutions lack efficiency, rigorousness or balance in design. An absence of a consistent book explaining how to analyse, design and develop new computer systems has been surprisingly revealed. That is why we attempt to introduce—as a first draft—a theory of resilient and evolving systems, as good as we see it today. We introduce rigorous concepts of system design with reconfigurability used when necessary for toleration of faults. Our concepts of redundancy have been theoretically justified and analysed. Redundancy of system we discuss taking into account technological aspects, including thermal barrier and reliability. We propose a new design of system and system software and describe hardware prototypes—to demonstrate feasibility of them. Simulations and trial runs are presented and explained as well.

This book at first was written for ourselves—for everyday work in safety-critical systems design. As it is ICT market and research in this domain are greatly segmented. Thus we had to create our own "meeting point" for all above-mentioned "customers" and addressing new properties of computer systems.

To "start all over again" researchers, engineers, users and even politicians should be ready to understand what future applications of ICT require, what kind of drawback of technologies we are facing, what are the limitations and how to find the most efficient structural solutions, accompanied by careful use of math methods.

This book is the first consistent work on our paradigm of evolving computers; it includes methods of analysis and synthesis of ICT with new properties such as evolving functioning, performance, reliability and energy-wise solutions. We also discuss abilities of system to match changing requirements and internal faults of hardware schemes, technological advances and drawbacks.

Initially this book did serve us an essential working material in terms of "all you need to know to design and analyze new generation of computer systems" addressing in non-mutually exclusive way reliability, fault tolerance, performance, resilience and properties of electronics, introducing supportive models and key hardware designs: processors, memories and interfaces. We were thinking about the following market our new system that developed, accordingly proposed approach: safety-critical, autonomous, real time, military, banking and wearable health care systems. Presented hardware prototype demonstrates at the order of magnitude higher efficiency in comparison with existing systems.

Who is our reader and why? Research community will get consistent area of further theoretical developments; Industries of hardware and system software designs, manufacturing and exploitation will get pathways to make performance, reliability and energy-smart systems with consistency, enabling unification of market of consumer electronics, safety-critical, embedded, autonomous and autonomous systems; Consumers will get much higher efficiency (and value for money) from their systems (if, of course devices and systems will be designed according to the principles proposed in the book).

This book provides also several personal benefits for the reader:

- Analysis of existing systems given in essence, showing how "classic" solutions stand and work.
- Existing technological drawbacks are clarified and presented consistently, with proposed solutions that "best fit the requirement" of new computer system.
- Description of a process of introduction of new properties as a framework required from next generation of computer system enabling a reader to make consistent analysis of—we stress—all possible system design solutions.
- Demonstrated and described prototype of evolving reconfigurable architecture might be attractive for students as they through the book will discover that computers might be designed much simpler, power efficient and at an order of magnitude more reliable.
- A prototype of the system and simulator will help for future engineers of embedded systems.
- Students and analysts will discover that the market dominance of the general computing systems has been now limited by appeared embedded systems with billions of units manufactured every year. Note that embedded systems appear in contexts where continuous operation is of utmost importance and failure can be profound.

– Any reader will be able to use trail simulator and start programming new architecture.

Nowadays radiation is a serious threat to the reliable operation of safety-critical systems. Fault avoidance techniques, such as radiation hardening, have been commonly used in space applications. However, hardened components are expensive, lag behind commercial components in performance and do not provide 100 % fault elimination. Without supportive structural solutions to provide fault tolerance, hardware faults become system errors at the application or system level, which in turn can result in catastrophic failures.

In this direction, we present known concepts of fault tolerance and dependability and extend them by our own concept of resilience and generalisation of fault tolerance. We propose to consider fault tolerance and resilience as *processes*, instead of properties. We analyse the physics of radiation-induced faults, the damage mechanisms of particles and the error as a consequence.

We propose new approach to hardware and system software design combining efficiently reliability, performance and power consumption.

Finally, to demonstrate how new properties of the computer system will be implemented, a new conceptual system element called a *syndrome* was introduced, described and its application for performance, reliability and energy-smart operations of hardware explained. Implemented by hardware and supported by system software *syndrome* serves as a core of a resilience of architecture enabling system (through software and hardware) be adaptable to various and modifiable functional requirements, different internal conditions and environmental impacts. We implemented a software simulator and disassembler and introduced a testing framework in combination with our evolving reconfigurable architecture assembler and commercial hardware simulators.

Stevenage, UK

Victor Castano
Igor Schagaev

Contents

Abbreviations

ARQ	Automatic repeat quest
ASIC	Application-specific integrated circuit
ASW	Application software
ATPG	Test pattern generation tools
BCH	Bose Chaudhuri hocquenghem
BEC	Backward error correction
BICMOS	Bipolar complementary metal oxide semiconductor
BIST	Built in self-test
BPSG	Boron phosphor silicate glass
CCD	Charged couples device
CED	Concurrent error detection
CM	Corrective maintenance
CMF	Common mode failure
CMOS	Complementary metal oxide semiconductor
COTS	Commercial off the shelf
CSP	Cold standby spare
CUT	Circuit under test
DDD	Displacement damage dose
DEC/TED	Double bit error correction and triple bit error detecting
DFT	Design for testability
DMR	Dual modular redundancy
DRAM	Dynamic random access memory
DRE	Detected recoverable error
DUE	Detected unrecoverable error
DUT	Device under test
DW	Data word
ECC	Error correcting codes
EDAC	Error detection and correction codes
EDC	Error detecting codes
EEPROM	Electrically erasable programmable read only memory

EPI	Epitaxial substrate doping
FCR	Fault containment region
FEC	Forward error correction
FIT	Failures in time
FM	Fault model
FPGA	Field programmable gate array
FT	Fault tolerance
FTS	Fault tolerant system
GAFT	Generalised algorithm of fault tolerance
GCR	Galactic cosmic ray
GDS	Gracefully degrading system
HARQ	Hybrid automatic repeat request
HSP	Hot standby spare
HW	Hardware
ICV	IDDQ checkable voter
IDDQ	Quiescent power supply currents
IDE	Integrated development environment
Iff	If and only if
IR	Information redundancy
LET	Linear energy transfer
MBU	Multiple bit upset
MCU	Multiple cell upset
MOS	Metal oxide semiconductor
MOSFET	Metal oxide silicon field effect transistor
MSB	Most significant bit
MTBF	Mean time between failures
MTTD	Mean time to detection
MTTF	Mean time to failure
MTTR	Mean time to repair/restore
NIEL	Non-ionising energy loss
nMMOS	n channel metal oxide semiconductor
NMR	N-modular redundancy
ORA	Output response analyser
PCSE	Power cycle soft errors
PDF	Probability density function
PI	Primary input
PKA	Primary knock-on atom
PM	Preventive maintenance
pMOS	p channel metal oxide semiconductor
PSF	Pattern sensitive fault
REDWC	Recomputing with comparison
RERO	Recomputing with rotated operands
RESO	Recomputing with shifted operands
RESWO	REDWC recomputing with swapped operands

RF	Register file
ROM	Read only memory
RS	Reed solomon
RT	Real time
RTS	Real time systems
SAF	Stuck at fault
SBU	Single bit upset
SDC	Silent data corruption
SEBO	Single event burnout
SEC DED	Single error correction and double error detection
SEDR	Single event dielectric rupture
SEE	Single event effect
SEFI	Single event functional interrupt
SEFLU	Single event fuse latch upset
SEGR	Single event gate rupture
SEHE	Single event hard error
SEL	Single event latch up
SEMU	Single event multiple upset
SER	Single event rate
SESB	Single event snapback
SET	Single event transient
SEU	Single event upset
SOC	System on a chip
SOI	Silicon on insulator
SOS	Silicon on sapphire
SPF	Single point of failure
SR	Structural redundancy
SRAM	Static random access memory
SSW	System software
SW	Software
TBF	Time between failures
TID	Total ionising dose
TMR	Triple modular redundancy
TMRV	TMR system with non-perfect single voting
TR	Time redundancy
TTF	Time to failure
TTR	Time to repair
UART	Universal asynchronous receiver/transmitter
WSP	Warm standby spares

Introduction

This book is the first consistent work on development of our paradigm of evolving computers; it includes methods of analysis and synthesis of evolving computer systems capable to modify the properties regarding performance, reliability or energy-wise functioning.

We also discuss abilities of system to match changing requirements and ability to tolerate internal hardware faults, with supportive analysis of technological advances and drawbacks.

Initially this book did serve us an essential working material in terms of "all you need to know to design and analyse new generation of computer systems" addressing in non-mutually exclusive way reliability, fault tolerance, performance and resilience properties of hardware. We have introduced supportive models for main hardware elements: processors, memories and interfaces.

We were thinking about the new architecture to fit the following market:

Safety-critical, autonomous, real time, military, banking terminal and wearable health care systems.

Proposed approach was developed up to hardware prototype—special chapters that describe it demonstrate at the order of magnitude higher efficiency in comparison with existing systems.

Who are our segments of society that might benefit from reading our book and why?

Research community will get consistent description of the theoretical area for further developments.

Industries of hardware and system software designs, manufacturing and exploitation will get pathways to make performance, reliability and energy-smart systems with consistency, enabling unification of market of consumer electronics, safety-critical, embedded, autonomous and autonomous systems.

ICT consumers will get much higher efficiency (and value for money) from their systems (if, of course, devices and systems will be designed according to the principles proposed in the book).

In turn, our book provides also several personal benefits for the reader in terms of:

- Analysis of existing systems given in essence, showing how "classic" solutions stand and work.
- Existing technological drawbacks are clarified and presented consistently, with proposed solutions that "best fit the requirement" of new computer system.
- Description of a process of introduction of new properties as a framework for next generation of computer system enabling a reader to make consistent analysis of—we stress—all possible system design solutions.
- Described prototype of evolving reconfigurable architecture might be attractive for students and hardware design engineers as they through the book will discover that computers might be designed much simpler, power efficient and at an order of magnitude more reliable, especially where continuous operation is of utmost importance and failure can be profound.
- Analysts who have discovered that the market dominance of the general computing systems has been now limited by appearance in billions of embedded systems will become aware how to merge these two market segments.
- Any reader will be able to use trail simulator and start programming for new architecture.

Nowadays radiation is becoming a serious threat to the reliable operation of safety-critical systems. Fault avoidance techniques, such as radiation hardening, have been commonly used in space applications. However, hardened components are expensive, lag behind commercial components in performance and do not provide 100 % fault elimination. Without supportive structural solutions to provide fault tolerance, hardware faults become system errors at the application or system level, which in turn can result in catastrophic failures. In this direction, we, along updated concepts of fault tolerance and dependability, introduce our own concept of resilience. To some extent resilience is a generalisation of fault tolerance. We propose to consider fault tolerance and resilience as processes, instead of static properties. We analyse the physics of radiation-induced faults, the damage mechanisms of particles and the error process. We propose new approach to hardware and system software design combining efficiently for reliability, performance and power consumption.

Finally, to demonstrate how new properties of the computer system will be implemented, we introduce a new conceptual system element called a syndrome. We describe how syndrome can be applied to monitor performance, reliability and energy-smart operations of hardware. Implemented by hardware and supported by system software, syndrome serves as a core of a resilience of architecture. Using syndrome enables the system (through software and hardware) to be adaptable to:

- various and modifiable functional requirements
- different internal conditions
- environmental impacts

We implemented a software simulator and disassembler and introduced a testing framework to our evolving reconfigurable architecture inviting researchers and engineers to put "hands on" new computer system with unique properties.

Stevenage, UK

Victor Castano
Igor Schagaev

Chapter 1
Basic Concepts, Motivation and Structure

1.1 Motivation

Embedded systems are ubiquitous nowadays, built into homes, offices, bridges, medical instruments, cars, airplanes and satellites and even into clothes. The market size of such systems is already larger than the one for general-purpose computing. The majority of embedded systems are real-time systems (RTSs) and mostly all RTSs are embedded either into device or product.

For decades, embedded RTSs are being used in fields where their correct operation is vital to ensure the safety and security of the public and the environment: from automotive systems and avionics to intensive health care and industrial control as well as military operations and defence systems. These systems are subject to time constraints and must guarantee a response within specified timing bounds. The safety critical nature of RT embedded systems employed in those fields demands the highest possible availability and reliability of system operation.

The technological achievements that have led to exponential growth of clock frequency and memory size have boosted the technological development of microprocessors. Manufacturers of advanced silicon electronics have been able to create more complex designs by periodically scaling down the technology, increasing the transistor density. This growth is supported by the progressive miniaturisation of electronic components predicted by Moore, called by technological bureaucrats as Moor's law in 1965.

In real world, technological developments have been slowed down by physical limitations: due to the reduction of size of electronic components to nanometer scales and due to the increase in clock frequencies (ITRS 2011), supply voltages have been reduced to keep power dissipation manageable while thermal noise voltages have increased (Asanovic et al. 2006; Kish 2002).

For a long time, environmental impact, radiation effects have been a serious concern in aviation, aerospace and special mission electronics. As the dimensions and voltages of embedded systems now are reduced, their sensitivity to ionising

V. Castano, I. Schagaev, *Resilient Computer System Design*,
DOI 10.1007/978-3-319-15069-7_1

particles has considerably increased. Energising particles can produce a number of faults at the hardware level, not only in contexts with harsh environmental conditions such as outer space but also at sea level with regular conditions.

Components with lower power and noise margins are less reliable, and therefore, recent systems are more prone to transient faults induced primarily by radiation (Baumann 2002, 2005a, b; Seifert et al. 2002; Shivakumar et al. 2002). Transient faults do not cause permanent damage in circuits but can affect system behaviour by corrupting stored information or signal communication (Karnik and Hazucha 2004; Mavis and Eaton 2002; JEDEC "JESD89-3A," 2007).

It is worth to point out that typical stress experiments in laboratory-based particle bombarding, present objective evidence of radiation induced malfunctions and catastrophic failures during operation in real life environments. Radiation-induced faults are frequent in space environments (Adams and Gelman 1984; Adams et al. 1982; Binder et al. 1975; Blake and Mandel 1986; Waskiewicz et al. 1986). The Saturn's Cassini (Swift and Guertin 2000), Deep Space 1 (Caldwell 1998), Mars Odyssey (Eckert 2001) and Jupiter's Galileo (Fieseler et al. 2002) are examples of missions that presented malfunctions as a result of cosmic rays. The satellites X-ray Timing Explorer (Barth et al. 2004), Gravity Probe B (Owens et al. 2006), TOPEX/Poseidon (Swift and John 1997) and GRACE (Pritchard et al. 2002) have also reported anomalies during operation.

Radiation induced faults are also present to a lesser extent in atmospheric (Taber and Normand 1993) and terrestrial environments (Hauge et al. 1996; Normand et al. 2010; Ziegler 1996).

As a result of this scenario there is an increasing need to deal with faults and their consequences in the system. There are two classes of mechanisms to cope with faults: *fault avoidance* and *fault tolerance* (FT) (Avizienis et al. 2004). *Fault avoidance* means developing components/systems that are less likely to present faults while fault tolerance techniques focus on the system's ability to tolerate the effects of these faults.

Fault tolerance is defined as *the ability to provide uninterrupted service, conforming to the desired levels of reliability even in the presence of faults* (Avizienis et al. 2004). Applications of modern electronic systems require much more powerful mechanisms to mitigate the effect of these faults (Nicolaidis 2010).

Complete avoidance of faults in a system is practically impossible and hence a balance of the two approaches is currently applied. With the objective of reducing costs, attempts to apply fault tolerance (FT) to COTS computers did not bring any substantial breakthrough in neither efficiency nor any other crucial property required from safety critical systems (Antola et al. 1986).

The research community mainly focuses on (a) identifying all possible mechanisms leading to accidents and (b) on providing pre-planned defence techniques against them. However, too little attention to potential systems that can respond to deviations from desirable states has been paid.

The research is driven by observations on resultant limitations from the evolution of computer architectures, which have been motivated by technological and market choices as well as physical limitations. These observations refer to performance deceleration, increase in power consumption/dissipation, reliability aspects, parallelization challenges, design complexities, and hardware and software inefficiencies. A brief explanation for each follows:

Performance decrease: As transistor density and frequency increase to satisfy the immediate market demands, unjustified complexity has been introduced in the current computer architectures. In recent years, clock rates of standard microprocessors have flattened and performance of processor cores has slowed down (Asanovic et al. 2006); (Franzon et al. 2010).

Power consumption/dissipation: Growth of processor clock speed increases power consumption and consequently power dissipation while die size remains the same. Therefore, the power/density ratio will keep increasing to the point where no practical technique can feasibly dissipate the generated heat.

Reliability: Performance, heat and power consumption are not the only concerns. Reliability of intra-chip communication is also affected by physical constraints. Transistor scaling shortens wire distances, thus improving performance, but it also implies thinning of those wires.

As wires become narrower, they also become taller in order to reduce the resistance per unit length. Media resistance limits the speed of electrons within. Long wires within close distance vary dependent timing characteristics at best and produce data corruption at worst. In short, thinner wires increase delays and harm reliability. Furthermore, as explained earlier, the same radiation fluxes that in the past had no effect on electronics are now able to induce faults that affect the logic value of current transistors with lower critical charge.

For these reasons, it is commonly believed by the research community that the classic Hardware/Software uniprocessor model has reached the power/performance-wall (Asanovic et al. 2006; Franzon et al. 2010).

Parallelization: The microprocessor industry approach is based on using the billions of transistors now available on a die to replicate the of-the-shelf core design multiple times and increase the size of caches. Nevertheless, effective programming of multi-core is not trivial and introduces multiple challenges (Geer 2007; Goth 2009; Pankratius et al. 2009). As an attempt to overcome the power wall, research on computer science has reincarnated parallel computing. Parallel computing and parallel programming are not new; they have been a mainstay in high-performance since the early 1950s (Hill and Rajwar 2001).

Complexity: The semiconductor industry, driven by economic reasons and time-to-market needs, has introduced unjustified complexity in microprocessor designs. An efficient and logical design could have achieved better results in the long-term. Instead, a brute force approach, increasing frequency and adding deeper pipelines and cache levels has been followed (Sager et al. 2001).

Software and Hardware Inefficiency: In terms of software, modular programming (Turski and Wasserman 1978; Wirth 1983) and later object oriented programming (Wirth 1988, 1992) were introduced to maximise performance and effectiveness of the human agent in the programming process.

To maximise performance of HW/SSW/ASW several approaches of parallelism using distributed, data-flow and cluster architectures were introduced in the late 1950s. The Flynn diagram (Flynn 1972) is still in use: SIMD (Single Instruction Multiple Data), MIMD (Multiple Instruction Multiple Data) and MISD (Multiple Instruction Single Data) are very well known architectures, each with their own benefits and drawbacks. In the early 1980s the VLIW (Very Long Instruction Word) (Fisher 1983) approach was also introduced. But since then, no significant new architecture has been introduced.

The size of the market is in the order of billions of controllers produced every year, which exemplifies the scale of dependency of human society on computer technologies. To make the next step in the design of special systems for safety critical applications we should analyse what is applicable from the well-developed theory and design of fault tolerant systems since early 1970s, in particular their reliability and resilience to electromagnetic impulses.

In turn, the success of future computer systems for safety critical applications will depend on trading-off performance, reliability and power consumption.

The combination of these two statements forms a framework for this work. At first, we should analyse the technological achievements of modern electronics in terms of performance. Finally, we should find ways to improve the efficiency of current embedded systems in terms of performance, reliability and power consumption.

1.2 Scope and Contribution

This work relate to the reliability of embedded systems with regards to permanent and transient hardware faults induced by radiation as the source of errors. Additionally, proposed methods techniques are also efficient to mitigate the effect of faults induced by other means. Note that software faults as the source of errors are out of the scope of our research. This section briefly explains our contributions.

The main goal of this research was to find efficient techniques and original mechanisms for improving reliability, performance and energy use of real-time systems in safety critical applications. Our special interest was in design, development and analysis of a fault tolerant reconfigurable architecture capable to tolerate radiation-induced faults. This is especially important as amount of induced faults is growing and without special efforts system architecture will degrade pretty fast. We believe and show it in a special chapter that such architecture will be further used as a core element for reconfigurable computer systems with key requirements of reliability, power awareness, performance and scalability.

We attempt to overcome known drawbacks of modern real-time systems. The outcomes of our efforts can be summarised as follows:

The traditional Reliability, Fault Tolerance and Dependability concepts and definitions do not take into account the transient nature of some of the faults induced by radiation. Therefore, a new concept of resilience has to be introduced reflecting the changing environment and the different FT contexts.

We examined systematically the physical mechanisms that lead to faults induced by radiation and the error process analysing and classifying the effects of radiation impact on modern microprocessor technologies.

We introduced a new model that classifies faults. This model enables and eases analysis of requirement during design process accounting serviceability and fault coverage, making fault tolerant and resilient system designs much more efficient.

We introduced a combination of structural hardware elements as three semantically different: active, passive and interfacing zones. In ensemble with system software, these hardware elements have to be developed differently making system resilience much more efficient and powerful. Combined hardware and system software design process that leads down to silicon technologies enables improving reliability and performance, as well as reducing power consumption making new designs far better that known systems.

We designed and implemented a hardware prototype as a proof-of-concept with core elements introduced in the previous statement.

Our hardware design work, development of simulation has benefited from collaboration and can be considered as a joint international research to make so-called Evolving Reconfigurable Architecture (ERA) with ETHZ Dr T. Kaegi, Dr E. Zoueff, Prof L. Blaeser published in Schagaev et al. (2010) and further developed in Schagaev et al. (2014).

We proposed a framework and testing scheme for the testing and debugging of the hardware prototype. As part of the framework, we implemented an assembler for the hardware prototype together with a disassembler and simulator tool.

1.3 Structure

We divide this work in ten chapters configured as follows:

Chapter 1: *Introduction* summarises the approach of fault tolerance and resilience, describes our contribution to this domain of science and defines general structure of further work.

Chapter 2: *Resilience*: presents elements of the theoretical framework for reliability. We analyse the properties of classic dependability and we describe our own view of the concept of resilience.

Chapter 3: *Dealing with faults*: *redundancy*: this chapter provides a complete review of state-of-the-art techniques employed to deal with faults and explores the different types of redundancy and fault tolerant techniques.

Chapter 4: *Impact of radiation in electronics of embedded systems* presents the physical mechanisms of radiation as the primary phenomenon that causes faults in current computing systems. We also analyse their effect on semiconductors at element, circuit and system levels.

Chapter 5: *Fault tolerance models* introduces our own model of hardware faults. We introduce generalised algorithm of fault tolerance (GAFT) and define the different states and actions required to implement fault tolerance.

Chapter 6: *Hardware support and System Software Support for Resilience* details the hardware and system software elements of a novel resilient architecture. Proposed architecture enables to achieve and change various levels of performance, reliability and energy consumption.

Chapter 7 describes software simulation aspects of hardware prototype, presents shortly a comparison of proposed architecture with known systems, describes benefit of simulation and use of simulator in the process of testing.

Chapter 8: *Implementation*: *Hardware Prototype, Simulation and Testing*: This chapter focuses on the development and testing of the hardware prototype for proposed hardware architecture. Details of the design and development of a software simulator of the hardware architecture are also provided.

Chapter 9 summarises our work, describes next steps in research and development, providing some hints of future of resilient systems theory, design and developments. Finally,

Chapter 10 presents our vision on Evolving Systems, fundamental constrains, supportive models, prototypes, challenges. Existing approaches are briefly compared with proposed evolving system concept.

Chapter 2
Background Concepts and Resilience

2.1 System Failure Life Cycle

Correct service (Laprie 1995) also named *proper service* (Laprie and Avizienis 1986) is delivered by a system when the service implements the function as specified. The fundamental threats to the correct service and to the resilience of safety critical systems are faults, errors and failures that, in turn, can cause catastrophic failures. Among these four terms there is a cause–effect relationship.

A *failure*, *service failure* or *system failure* is an event that takes place when the delivered service deviates from proper service. Hence, a service failure implies a transition of the system from proper service to an improper service, not implementing the functions as specified by the functional specification of the system.

The downtime or *period of delivery of improper service* is also referred to as *service outage*. *The transition* from *improper service* to *proper service* is called *service restoration*, *service recovery* or *repair*.

Since a service is an organised sequence of the internal and external states of a system, a service failure takes place when one or more of its external states deviate (s) from the correct service state. These deviations are *errors*. An *error* is a part of the system state that is liable to lead to a subsequent *failure*. The hypothesised or adjudged cause of such error is a fault.

A *fault* is a weakness, blemish or shortcoming of a particular hardware component or unit.

An *error* is the manifestation of a fault, a deviation from accuracy or correctness. Finally, if the error leads to one of the system's functions being performed incorrectly then *a failure* has occurred.

Figure 2.1 graphically describes the well-known life cycle of system failure within a three universe model (Johnson 1989) adapted from the four universal model originally developed by Avizienis (1982). In the first universe, the physical

V. Castano, I. Schagaev, *Resilient Computer System Design*,
DOI 10.1007/978-3-319-15069-7_2

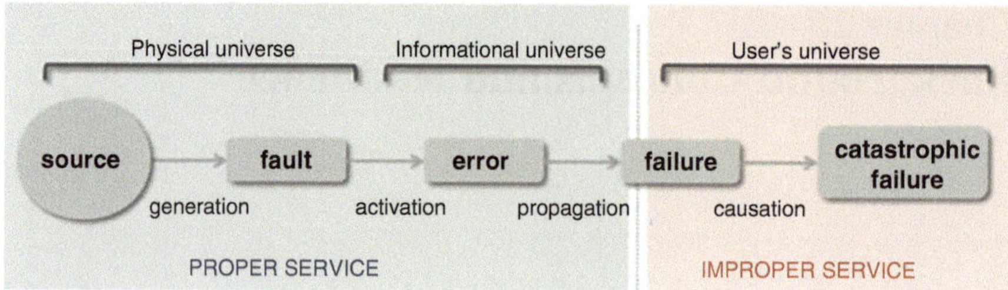

Fig. 2.1 System failure life cycle within a three-universe model

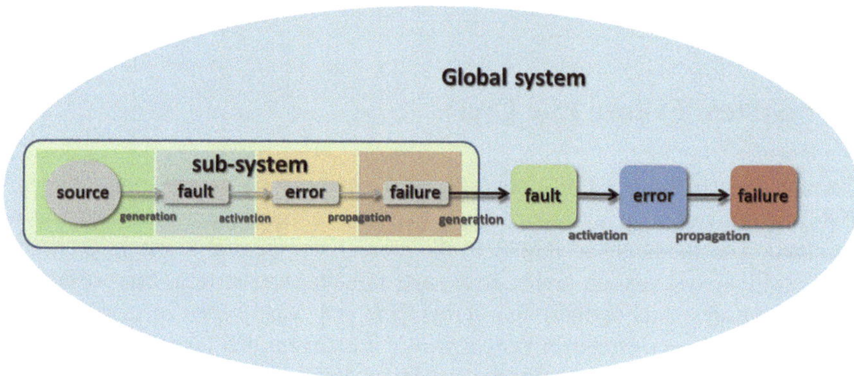

Fig. 2.2 Fault–failure: Transition between different levels of a system

one, faults are generated due to various sources. Faults can activate errors within the second universe, the informational one.

Errors take place when some information units become incorrect. In turn, errors could propagate the user universe and lead to a failure. It is in this final universe, where the user can witness the effects of faults and errors in the form of failures. One or more failures could potentially cause a catastrophic failure in the case of safety critical systems.

The arrows between the entities in Fig. 2.1 correspond to latencies. Fault latency (activation latency in Fig. 2.1) is the period of time between the occurrence of a physical fault and the appearance of an error. Likewise, error latency is the length of the propagation time that takes place between the activation of the error and the manifestation of the failure.

The term *fault* and *failure* are sometimes unclear in reliability literature. In this work, the term *fault* is sometimes equivalent to *failure*. For instance, a *system failure* can be the same as a component fault; this with a growing density of electronics components is a physically proven truth. Figure 2.2 shows the *fault–failure* transition between a subsystem and a global system.

The *fault–failure* cycle can be applied at different levels of abstraction within a system; consider a transistor as a subsystem that is part of a more global system (e.g. memory cell): the occurrence of incorrect functionality of the transistor during normal operation (e.g. the effects of ageing and stress) is a subsystem failure of such component but may lead to, for instance, a logic fault (*global system failure*).

This logic fault will remain dormant unless it is activated, producing an error, which is likely to propagate and create other errors. If the correct service of that global system is affected, a global system failure occurs. The same subsystem–system transition can take place between the memory cell, the memory circuit that the cell is part of, the microprocessor system that can be part of a multiprocessor, etc.

2.2 Attributes and Measures of Resilience

The word *resilience* from the Latin etymology *resilire* (to jump back, or to rebound) is literally the tendency, ability, act or action of springing back, and thus the ability of a body to recover its normal shape and size after being pushed or pulled out of shape. That is, the ability to recover to normality after a disturbance, shock or deviation from the intended state and go back to a pre-existing or acceptable or desirable state.

The meaning of resilience is somewhat different between authors. The term *Resilient* has been traditionally used as a synonym of fault tolerance (Laprie 2008). Before we discuss fault tolerance as a concept and introduce resilience, several other terms need to be defined.

One of them is *Dependability*, which is an integrative concept that encompasses many other quantitative and qualitative attributes. Laprie (Laprie et al. 1992) defines dependability as the "trustworthiness of a computer system such that reliance can be justifiably placed on the service that it delivers".

Dependability is the ability to deliver a service that can justifiably be trusted. Laprie defines the service delivered by the system as its behaviour as it is perceptible by its users); a user is another system physical or human) which interacts with the former. Such service is classified as "proper" or "correct" if it is delivered as specified; otherwise it is considered as "improper" or "incorrect" (Laprie and Avizienis 1986). Again, the "properness" or "correctness" of the system service depends on the viewpoint of the user.

The terms related to *dependability* have been redefined over the years (Avizienis et al. 2004). In order to make the discussion unambiguous, and avoid conflicts of terminology due to different opinions of various authors, we merge and organise the attributes or measures of dependability and adapt them to the field of safety-critical applications. The attributes of dependability are reliability, safety, performability and security. The latter encloses a subset of attributes including integrity, maintainability and availability.

2.3 Reliability

The reliability measure is most often used to characterise systems in which failures are unacceptable, therefore including field of safety critical systems.

Figure 2.3 shows a non-repairable system with two possible states: a fully functional start state (up) and a failed state (down), involving loss of functionality, which can be reached after a transition due to failure.

There is no disagreement about the need for reliable systems but some vague notion of reliability is not enough in safety-critical engineering. Reliability can be defined as follows:

Reliability R(t) is the probability that a system or component will perform its intended function (without failure) over the entire interval [0,t] under specified environmental and operating conditions.

2.3.1 Performance and Reliability

2.3.1.1 Power-Reliability Wall

Since the invention of the integrated circuit in 1958 each generation of semiconductor technology has exponentially decreased the transistor price and exponentially increased the transistor density per chip (Hutcheson 2009).

This technological shrink has led to the impressive level of technology and hardware element density relatively recently achieved (Nair 2002) with processor frequencies reaching up to 4.7 GHz Power 6 Specs: IBM Power6 Microprocessor and IBM System p 570, 2007. The higher number of transistors and the kilometres of wire operating at higher frequencies requires up to several Watt/cm^2 of energy for on modern chips, leading the peak energy consumption well over 140 W. Most of that energy becomes heat, rising operating temperatures.

Heat and power consumption are not the only concerns of modern electronics. Reliability and performance of intra-chip communication are also affected by physical constraints. Transistor scaling shortens wire distances, improving performance, but it also implies thinning of those wires. The resistance and capacitance of the media limit the speed at which electrons can flow. Thinner wires increase delays and harm reliability (Shivakumar et al. 2002). For these reasons, the research

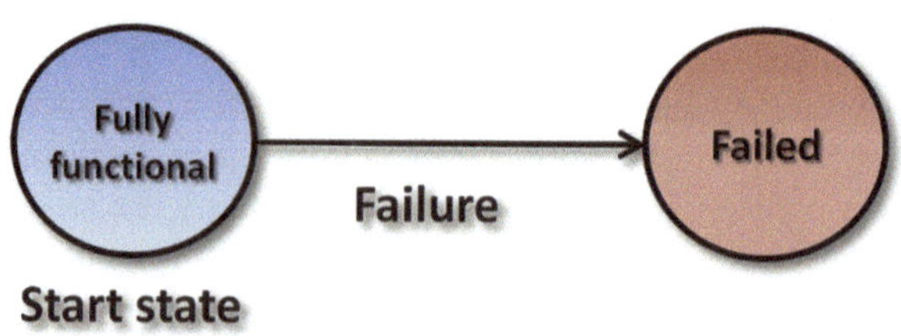

Fig. 2.3 A non-repairable system with two states

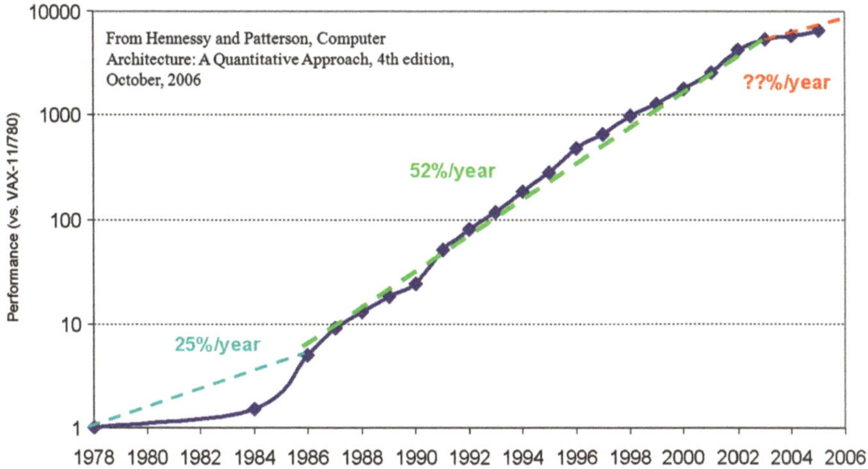

Fig. 2.4 Growth of performance since the mid-1980s (Hennessy and Patterson 2006)

community believes that the classic HW/SW single processor model has reached the power-reliability wall (Asanovic et al. 2006).

Moreover, the cost of technologies used for the manufacturing is increasing exponentially. Such cost is doubling every 4 years, which makes smaller nanometer-scale technologies required; pursuing these efforts presents not only technical but an economic challenge as well.

Evidence of this phenomenon is the chart shown in Fig. 2.4 that plots performance gap between the processor and memory of the VAX 11/780 measured by the SPECint benchmarks[1]). Subsequent to the mid-1980s, processor performance growth averaged about 52 % per year. Since 2002, uniprocessor performance has slowed down to about 20 % per year reaching the power-reliability wall in 2006. On the other hand, memory has averaged a constant performance increase of 9 %.

Using so-called Moore's law as a metric of progress has become misleading, as improvements in transistor density no longer translate into performance and energy efficiency. Starting around the 65 nm technologies, transistor scaling no longer delivers the performance and energy gains that drove the semiconductor growth during the past decades (Dreslinski et al. 2010).

The research community and the industry believe that parallelism is the answer to overcome the performance wall; however, some debates take place about different approaches of parallelism support and implementation. The industry has attempted to react by escalating the number of processors introducing multi-core architectures and parallelism.

Multiplying the number of big, complex and power demanding existing cores, which are part of the problem, does not adequately provide or resolve any of our

[1] SPECint benchmarks are a set of benchmarks design to test the integer processing performance of modern CPU http://www.spec.org/).

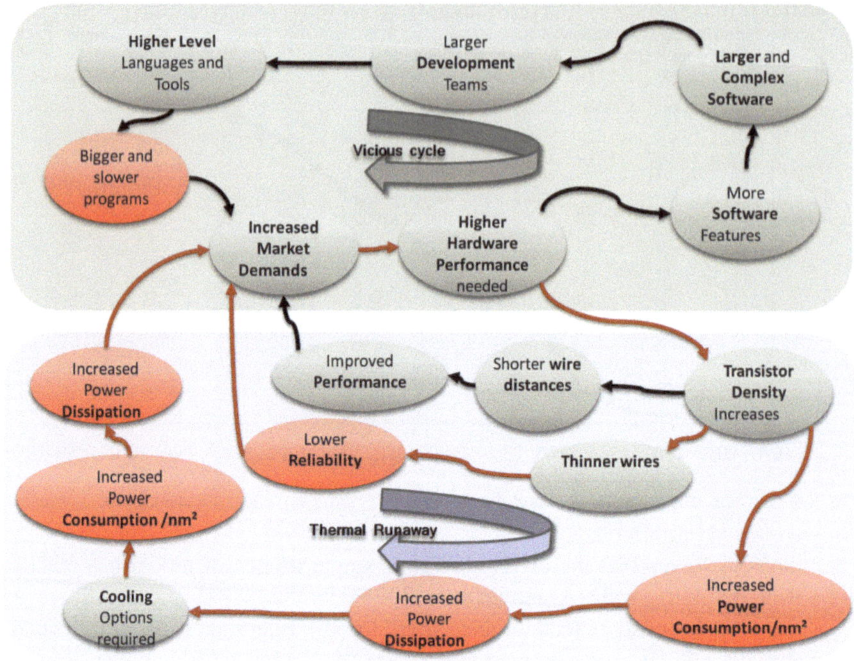

Fig. 2.5 The vicious cycle and the evolution of computing systems, 1950–2005

performance problems, but add concerns for reliability and power consumption (Asanovic et al. 2006). Replicating the essence of the problem will hardly add a solution to it.

2.3.1.2 Reliability with Vicious Cycle

The semiconductor industry is driven by competition, economic conditions and time-to-market. All this leads to an introduction of too much complexity in microprocessor designs. Figure 2.5 shows our interpretation of the reliability problem in current computing.

An efficient and logical design and redesign could have achieved better results in the long term. Instead, the industry has employed a "brute force" approach, by increasing hardware frequency, making larger and deeper pipelines and creating extra cache memory levels. All this did provide a slight performance gain at a high cost of chip floor plan, as shown by Asanovic et al. (2006).

Increased processor performance allows software companies to develop larger and feature-richer software, which involves larger development teams. Consequently, developers need higher-level languages and abstractions, which are less efficient and generate slower programs. As a result, again, faster processors are required, reinforcing the same vicious cycle (Fig. 2.5) and generating detrimental results. Pursuing this, existing programs would run faster on the latest generation of microprocessors, with performance gain at max 12 % with multiplied cost and energy consumption of hardware.

Since 2005–2006, there is no evidence of any considerable increase in functional hardware performance. Existing programs need to be redeveloped to take advantage of new multi-core. Consequently, the vicious cycle does not apply straightforward anymore. The power wall has dramatically slowed down the evolution of microprocessors in terms of performance and reliability.

Clearly, technological developments have not been supported by a logical evolution. There is an increasing need for unified hardware and software technologies. Rigorously speaking a new computing paradigm is required with its implementation through the whole cycle of hardware, software and application design as well as development and prototyping.

2.3.2 Reliability and Unreliability Functions

Let us suppose a system with N identical components. We define $S(t)$ as the number of surviving components at time t and $Q(t)$ as the number of failed components up to time t. Therefore, Eq. (2.1) has a form:

$$S(t) + Q(t) = N \tag{2.1}$$

The reliability $R(t)$ is the proportion of components that continue to perform without failure after being used for a period of time t. That is the probability of survival of the components, given by Eq. (2.2):

$$R(t) = \frac{S(t)}{N} \tag{2.2}$$

Unreliability, or *Cumulative failure distribution function* is generally referred to as the *probability of failure*. More specifically, *unreliability* $F(t)$ is the *conditional probability* that the system begins to perform incorrectly during the interval $[t_0,+t]$, given that the system was performing correctly at time t_0 Eq. (2.3):

$$F(t) = \frac{Q(t)}{N} \tag{2.3}$$

Based on Eq. (2.1) reliability and probability of failure of components at time t:

$$\begin{aligned} R(t) + F(t) &= 1 \text{ and} \\ F(t) &= 1 - R(t) \end{aligned} \tag{2.4}$$

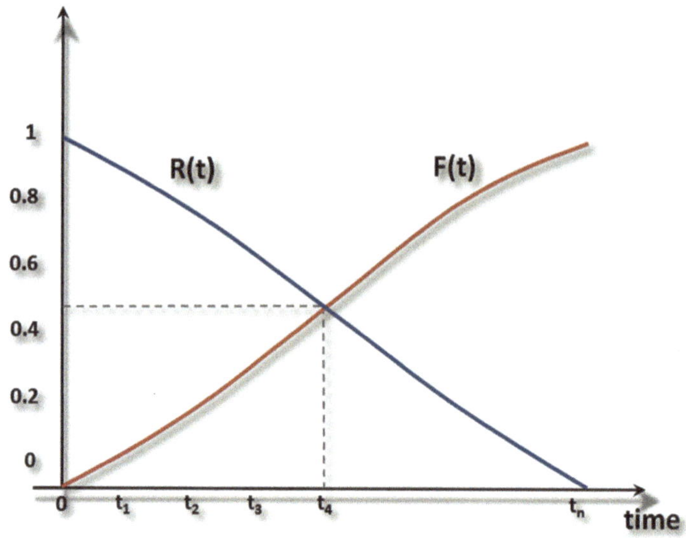

Fig. 2.6 Reliability $R(t)$ and Probability of failure $F(t)$ functions over time t

In turn, Fig. 2.6 presents a graph of the reliability and probabilities of failure over time with a constant failure rate. $R(t)$ is a monotonically decreasing function that has an initial value 1, whereas $F(t)$, starting at 0, increases monotonically. The sum of $F(t)$ and $R(t)$ at any given time is 1.

2.3.3 *Probability Density Function*

The derivative of $F(t)$ is a probability distribution function (PDF) that defines the probability of failures per unit time $f(t)$ of a particular component that has been used for a period of time t. Based on this definition, the probability density function is described as Eq. (2.5):

$$f(t) = \frac{\mathrm{d}F(t)}{\mathrm{d}t} \tag{2.5}$$

Using Eq. (2.4):

$$f(t) = \frac{\mathrm{d}[1 - R(t)]}{\mathrm{d}t} = -\frac{\mathrm{d}R(t)}{\mathrm{d}t} \tag{2.6}$$

Thus, the probability of a failure during the time range $[0,t]$ is Eq. (2.7):

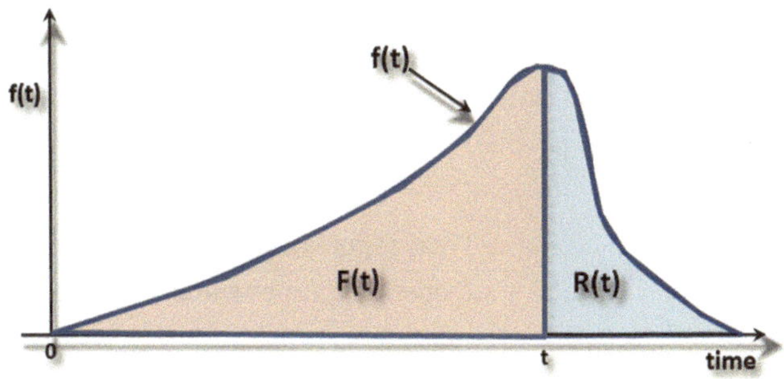

Fig. 2.7 Representation of reliability, unreliability and the probability density function

$$F(t) = \int_0^t f(t)\mathrm{d}t \tag{2.7}$$

Again, using Eq. (2.4) for Reliability over range [0, t] Eq. (2.8):

$$R(t) = 1 - F(t) = 1 - \int_0^t f(t)\mathrm{d}t \quad = \int_t^\infty f(t)\mathrm{d}t \tag{2.8}$$

Schematically, Fig. 2.7 presents Reliability, Unreliability (left and right areas, respectively) and PDF.

2.3.4 *Failure Rate Function*

The failure rate function $\lambda(t)$ also known as momentary failure rate or hazard function describes the number of failures per unit of time versus the number of components still operating at a time surviving components Eq. (2.9):

$$\lambda(t) = \frac{1}{S(t)} \frac{-\mathrm{d}Q(t)}{\mathrm{d}t} \tag{2.9}$$

Using Eqs. (2.3) and (2.2), failure rate as a function of reliability and probability density will have a form Eq. (2.10):

$$\begin{aligned} \lambda(t) &= \frac{1}{N\,R(t)} \frac{N\,\mathrm{d}F(t)}{\mathrm{d}t} \\ \lambda(t) &= \frac{1}{R(t)} \frac{\mathrm{d}F(t)}{\mathrm{d}t} = \frac{f(t)}{R(t)} \end{aligned} \tag{2.10}$$

The failure rate function is useful for analysis of reliability when semiconductor components operate with long period, or life cycle of the device.

However, calculating the failure rate at a specific point of time within a short period is impractical. Consequently, average failure rate, with longer time periods, is preferred Eq. (2.11):

$$\text{Average failure rate} = \frac{\text{Total failures during a period}}{\text{total operating time within a period}} \tag{2.11}$$

The values of average failure rate can be expressed by % or ppm.[2] However, FIT[3] is more widely used as a unit for reliability.

2.3.5 Cumulative Hazard Function

Using Eq. (2.4), failure rate as a function of reliability can be presented as:

$$\lambda(t) = \frac{-1}{R(t)} \frac{\mathrm{d}R(t)}{\mathrm{d}t} \tag{2.12}$$

The limits of integration can be obtained as follows:

- At time $t = 0$, $R(t) = 1$
- At time t by definition the reliability is $R(t)$

Given the assumption of a constant failure rate λ of a component (typically in per million hours or FIT):

$$\begin{aligned} \lambda t &= -\log R(t) \\ -\lambda t &= -\log R(t) \\ R(t) &= \mathrm{e}^{-\lambda t} \end{aligned} \tag{2.14}$$

2.3.6 Bathtub Curve of Failure Rates

In the world of technological requirements a reliability of a system is measured with values either from 0 to 1 or in %. Thus, a system may have 97 % reliability over a 2-year mission, subject to a maximum vibration Vmax, a humidity range [Hmin, Hmax] and a temperature range [15 °C, +30 °C].

[2] ppm is the abbreviation of "parts per million". One ppm means 1 faulty component out of 1,000,000 components. Hence, an average failure rate of 10 ppm means that there are 100 faulty components out of 1,000,000, or 1 component out of 100,000.

[3] FIT is a unit widely used to express failure rate. One FIT equals to one failure per billion (10^9) hours (one failure in about 114,155 years), or 1 ppm/1,000 h.

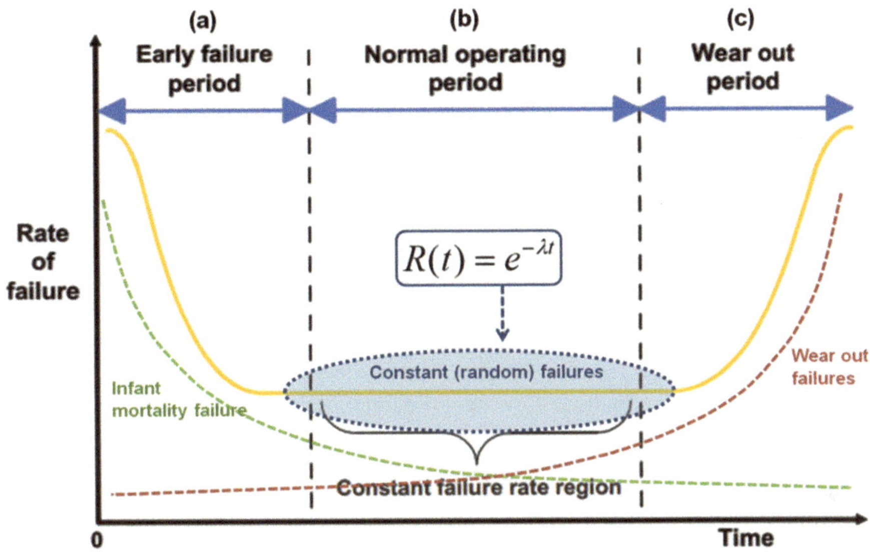

Fig. 2.8 A bathtub curve of failure rates. During normal operation period the failure rate λ is constant and faults are independent

Although the above definition is generally accepted, it is not really a complete one as from the start of operation till the end of device life system reliability will vary. Therefore, more factors need to be considered. For a correct service delivery in a specific period, the system must be operating properly at the beginning of the observation period.

The age of the system is one of the factors that should be taken into account. The above definition does not differentiate between:

- A new system.
- A system that has been operational for a substantial amount of time and whose faults have already been corrected.
- An old system with a long operational history and wear out issues.

Reliability distributions with decreasing, constant and increasing failure rate as a function of time are illustrated in Figure 2.8 separating periods (a), (b) and (c), respectively. The assumption made is that faults are independent and that the failure rate (λ) is constant.

The system failure rate is dependent on the system's lifetime constituting a function with a bathtub shape and three distinctive areas or periods: an early failure period (a), a normal operating period (b) and a wear out period (c). For failure rates higher than the constant failure rate (λ), the chance of system failure becomes higher.

For a new system (segment a) there is an early failure or infant period with a decreasing but high failure rate due to latent manufacturing defects that escape the initial testing of the product. As the products get into operation, these defects surface quickly when the devices are stressed.

Once the infant failures are eliminated, this high failure rate rapidly decreases to an almost constant value during the normal + operating or grace period (segment b). This long period represents the useful life of the system where failures occasionally occur due to the sporadic breakdown of weak components. It is highly desirable that this period of low failure rate and high reliability dominates the product's lifetime.

During the wear out or breakdown period (segment c) the reverse situation takes place. As the system gets older, the failure rate increases sharply due to age-related wear out. Note that many devices that form part of the same system will initiate this phase roughly at the same time. This could create an avalanche effect that could critically decrease the overall reliability of the system.

After analysing the bathtub curve and the three periods of operation involved, it is clear that the previous equations of reliability only suit the normal operating period with a constant failure rate. This curve is used for reliability analysis of hardware reliability reflecting processes of ageing and degradation.

At the same time similar reliability assumptions and models do not suit reliability analysis of software, especially when versioning and upgrading of software take place. The physics of hardware failure is discussed in Chap. 4, where properties of silicon substrates and radioactive particles are taken into account.

2.3.7 Mean Time Between Failures (MTBF)

Instead of a monotonic function of time a system reliability can also be expressed as a single numeric index. Mean time between failures (MTBF) is the average time that the system will run between failures. This measure is convenient to compare the reliability of different repairable systems. MTBF can be estimated by averaging the time between failures, including any additional time required to repair the system and place it back to a functional state.

Being $f(t)$ the probability of failure per unit time, MTBF can be described by Eq. (2.15):

$$\mathrm{MTBF} = \int_0^\infty t f(t) \mathrm{d}t \tag{2.15}$$

Using Eq. (2.6) then MTBF is Eq. (2.16):

$$\mathrm{MTBF} = -\int_0^\infty t \frac{\mathrm{d}R(t)}{\mathrm{d}t} \tag{2.16}$$

Integrating the above equations by parts we obtain:

$$\mathrm{MTBF} = -[tR(t)]_0^\infty \int_0^\infty R(t)\mathrm{d}t \tag{2.17}$$

For $t = 0$, $R(t) = 0$, hence $t \times R(t) = 0$. As t increases from 0, $R(t)$ decreases. As t tends to ∞, $t \times R(t)$ tends to zero. Therefore, the first term of the previous equation is zero. For any kind of failure distribution with a failure rate λ as a function of time, the general expression for MTBF can be described by Eq. (2.18):

$$\mathrm{MTBF} = \int_0^\infty R(t)\mathrm{d}t \tag{2.18}$$

The higher the MTBF, the higher the reliability of the system or component. Moreover, for failure distributions independent of time with a constant rate, MTBF is given by Eq. (2.19):

$$\mathrm{MTBF} = \int_0^\infty \mathrm{e}^{-\lambda t}\mathrm{d}t \tag{2.19}$$

and

$$\mathrm{MTBF} = \frac{1}{\lambda}\left[\mathrm{e}^{-\lambda t}\mathrm{d}t\right]_0^\infty = \frac{1}{\lambda} \tag{2.20}$$

Hence, MTBF of a system is reciprocal to its failure rate (given a constant failure rate). MTBF will be expressed in hours if the constant rate is also expressed in hours.

2.3.8 Mean Time to Failure (MTTF)

As described above, MTBF is a good measure of reliability for systems that can be repaired. A similar single-parameter indicator of reliability for components that cannot be repaired is the mean time to failure (MTTF). MTTF is the average time until the first system's failure.

Results of life testing can be used to calculate MTTF by testing a set of N identical units until all of them have failed with the time to the first failure of the individual units identified as $t_1, t_2, t_3, \ldots, t_n$. It can be observed that MTTF is given by Eq. (2.21):

$$\mathrm{MTTF} = \frac{1}{\lambda}\sum_{i=1}^{n} t_i \tag{2.21}$$

As before, the failure rate, if independent of time, can be calculated by Eq. (2.22):

$$\lambda = \frac{1}{\mathrm{MTTF}} \tag{2.22}$$

MTBF and MTTF are sometime used interchangeably. Although the numerical difference is small in many cases, both measures represent different concepts. MTTF is related to MTBF but does not include the repair time (MTTR or mean time to repair/restore), nor does it include the detection time (MTTD or mean time to detection) (Eq. (2.23)):

$$\mathrm{MTBF} = \mathrm{MTTF} + \mathrm{MTTD} + \mathrm{MTTR} \tag{2.23}$$

MTTR is the average time required to repair a system, whereas MTTD is the average time required to detect a failure. In most applications, MTTR and MTTD are just a small fraction of the total MTTF. Therefore, the approximation that MTBF and MTTF are almost equal is sometimes fair. MTTR and MTTD are difficult to estimate and can be determined by injecting faults into a system, measuring the time required to repair it. Both measures are further discussed in the availability section.

2.3.9 Reliability Prediction

In the case of design of hardware systems, there are two different known theoretical methods to meet the above mentioned reliability requirements and specifications:

- Fault avoidance: makes use of substantially higher reliability components and substantially higher than expected lifetime. Birolini (2014) introduced a comprehensive theoretical approach based on the application of reliability engineering throughout the system to reach this goal.
- Fault tolerance: deliberately introduces redundancy in the system to achieve continuous operation.

During the last 50 years there have been several attempts (Gnedenko et al. 1999; Koren and Krishna 2007; Kovalenko et al. 1997; Birolini 2014) to connect probability and reliability. A brief review of the application of probability theory for the analysis of reliability of real objects and their features (fault tolerance) is presented below.

Reliability of systems can be estimated by partitioning those systems into more elemental entities (e.g. subsystems or components). This partitioning enables to assess the individual reliability of these individual entities. The entities can be interconnected in serial, parallel or be considered as mixture of both. Therefore, reliability models are needed to estimate the role and involvement of entities or elements in the reliability of the system. Then we are able to analyse how a failure of each component would affect the overall reliability of the system.

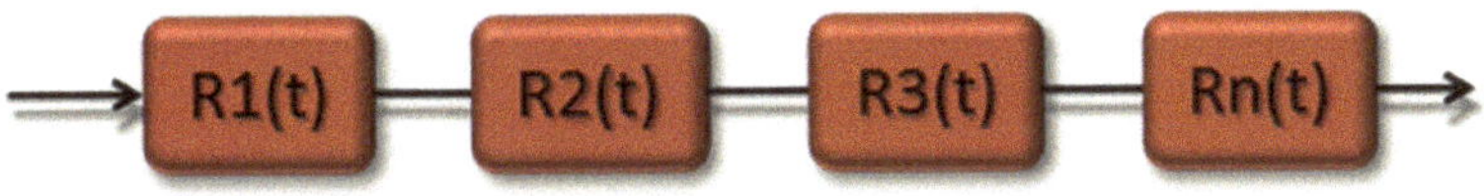

Fig. 2.9 Logic diagram of serial system reliability

2.3.9.1 Serial Reliability

In this model, the entities are connected in series. When minimum design and costs are specified in the design requirements of a system, a series system is the usual choice for designers. For the system to be operational, all of the components or subsystems should be operational and work correctly.

Serial systems are inherently unreliable since the failure in one of the elements would cause a stoppage of the overall system. The reliability of a system without redundancy may be described with a sequential reliability block diagram (see Fig. 2.9).

In this arrangement the system reliability is the product of its individual component reliabilities, assuming they are organised in serial (cumulative) structure. Note that for this structure, if the reliability of each component is R_i, the total system reliability R_s is given by Eq. (2.24):

$$R_s(t) = \prod_{i=1}^{n} R_i(t) = \exp\left(-\left(\sum_{j=1}^{n} \lambda_j\right)t\right) \tag{2.24}$$

And the failure rate of a serial system λ_s is given by Eq. (2.25):

$$\lambda_s = \lambda_1 + \lambda_2 + \lambda_3 + \ \ldots\ + \lambda_n \tag{2.25}$$

Furthermore, the Mission Time Function MT(r) gives the time at which system reliability falls below the given threshold level r. The relationship between reliability $R(t)$ and mission time MT(r) is given by the definitions:

Mission time function MT with threshold level r (Eq. (2.26)):

$$R[MT(r)] = r \tag{2.26}$$

Mission time function MT at given time t (Eq. (2.27)):

$$MT[R(t)] = t \tag{2.27}$$

If λ is constant then, using Eq. (2.14) then Mission time function with constant failure rate (Eq. (2.28)):

$$\begin{aligned} t &= \frac{-\ln(r)}{\lambda} \\ MT(r) &= \frac{-\ln(r)}{\lambda} \end{aligned} \tag{2.28}$$

Respectively, for a non-redundant system with n components Mission time function for non-redundant system with n component (Eq. (2.29)):

$$MT(r) = \frac{-\ln(r)}{\sum_{i=1}^{n} \lambda_i} \tag{2.29}$$

The failure rate of a sequential independent element system is equal to the sum of the failure rates of its elements. In the case of a constant failure rate across all elements, the MTTF of the whole system (MTTFS) can be calculated as follows (Eq. (2.30)):

$$\mathrm{MTTF_S} = 1/\lambda_s \tag{2.30}$$

Note that this equation highlights the fact that the reliability of a system is directly impacted (in practice often dominated but not solely determined) by the reliability of its least reliable component.

2.3.9.2 Parallel Systems Reliability: Redundancy and Fault Tolerance

In the previous model, no redundancy was taken into account to calculate the system reliability. A second approach to achieve a required level of reliability is the deliberate introduction of extra components into the system. The sole purpose of introducing this redundancy artificially is to increase reliability. However, there is a price to pay for such improvement in the system's reliability.

This approach assumes a deliberate introduction of redundancy in the system and has been applied since the original work of Von Neumann (1956) and Pierce (1965). Note that introducing redundancy involves some additional components and complexity, and it is therefore imperative that the reliability benefit accruing from the redundancy scheme must far exceed the decrease in reliability due to the actual implementation of the redundancy mechanism itself.

The classic parallel generalisation of the redundancy model (Birolini 2014) describes a system of n statistically identical elements in active redundancy, where k element(s) is/are required to perform a function and the remaining n–k are in reserve.

A function of the system is considered successful if during scheduled time k element(s) of the system was/were available. As an example, in the case of a 1-out-of-3 system (Fig. 2.10), its function would be complete if at least one of the elements was known to be working correctly. The second and third elements are redundant and introduced only for reliability purposes when the first unit is known to be faulty.

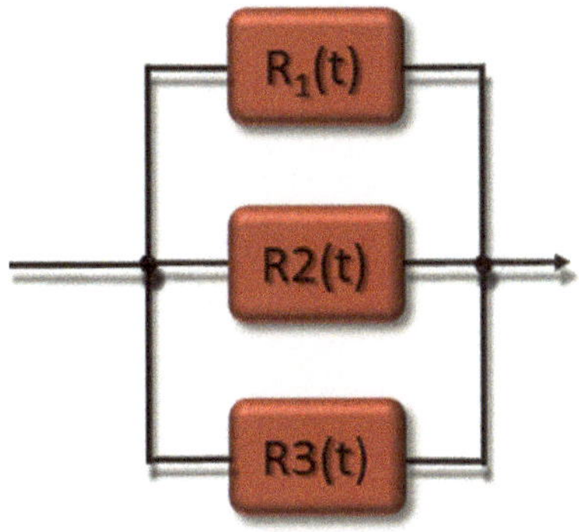

Fig. 2.10 Parallel reliability

For the system of Fig. 2.10 the reliability function is as follows:

$$R(t) = R_1(t) + R_2(t) + R_3(t) - R_1(t)R_2(t)R_3(t) \quad (2.31)$$

Assuming that the elements are identical, work or fail independently of each other and have constant failure rate $R_1(t) = R_2(t) = e^{-\lambda t}$, then by substitution:

$$R(t) = 3e^{-\lambda t} - e^{-3\lambda t} \quad (2.32)$$

And mean-time to failure (Eq. (2.33)):

$$\mathrm{MTTF_s} = \frac{3}{\lambda} - \left(\frac{1}{3}\right)\lambda = \frac{8}{3\lambda} \quad (2.33)$$

Therefore, the apparent working time of the redundant system is increased. In the general case where n redundant elements are introduced as spares to provide successful completion of an element's function with the same assumptions as above, the overall system reliability is given by Eq. (2.34):

$$R(t) = 1 - \left(1 - e^{-\lambda t}\right)^n \quad (2.34)$$

In the above equation n is the number of modules, $e^{-\lambda t}$ is the reliability of the original system and it is assumed that:

- There is a fault-free mechanism to detect and report failure of the active module.
- There is a fault-free switching mechanism to replace the active module in case of detected failure.
- All modules have equal reliability.

Thus, there is no doubt that redundancy even for this classic case could improve reliability of the system considerably. Note that the redundant components do not necessarily need to be identical, but could also correspond to additional hardware with different reliabilities used to detect and treat transient faults.

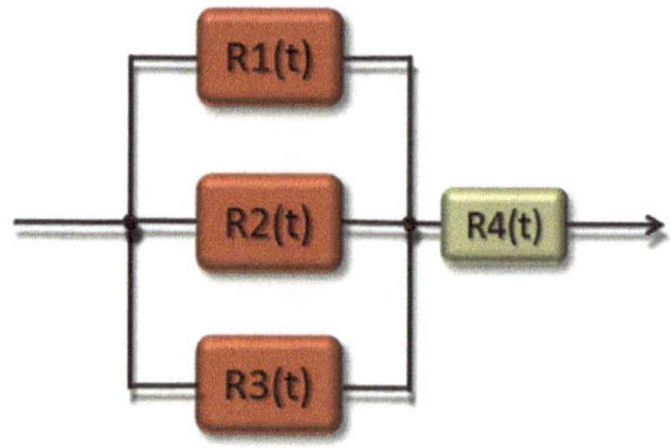

Fig. 2.11 Reliability of a combination of serial/ parallel components with a voter

2.3.9.3 Reliability for Mixed Serial and Parallel System

In practice, systems are usually made of a combination of serial and parallel components. More complex math applies to the reliability of these mixed arrangements. This type of arrangement is frequently used in systems where a specific part is particularly prone to failure. Figure 2.11 depicts an example of M of N system, whose elements may or may not have constant rates, and has a voter that counts for the serial reliability element.

Assuming that only 1 out of N parallel components needs to operate, the reliability of the parallel section of the system is defined by Eq. (2.35):

$$R(t) = 1 - [(1 - R_1(t))(1 - R_2(t))(1 - R_3(t))] \tag{2.35}$$

The total reliability of the mixed serial/parallel system shown in Fig. 2.11 is specified by Eq. (2.36):

$$R(t) = R_{1-3}(t)R_4(t) \tag{2.36}$$

Therefore, a relatively reliable voter would dominate the reliability of a redundant system.

2.4 Safety

In safety critical systems, safety describes the absence of catastrophic failures for users and the environment when a failure takes place. A system that can be repaired after failure presents a minimum of two states: functional and failed. Some other systems are able to have extra states even under faulty conditions. An example of such system, depicted in Fig. 2.12, has the possibility of transiting to a safe state, in a manner that does not cause any harm.

Safety is a measure of the fail-safe capability of a system, and it is defined as the probability that a system will either perform its function correctly or will

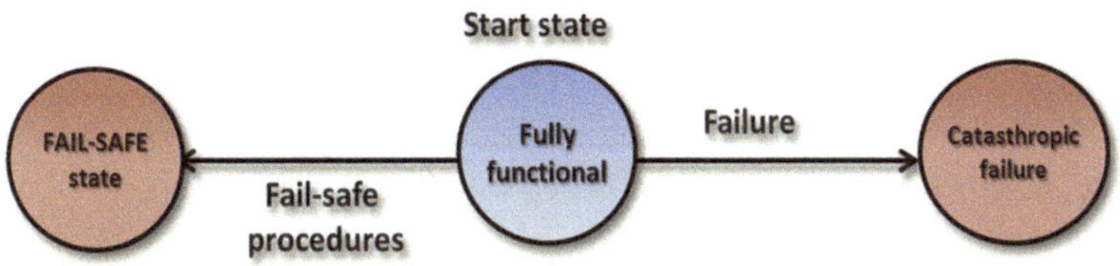

Fig. 2.12 A basic fail-safe system with three states

discontinue its operation in a safe way. Quantitatively, safety is the probability that the system will not fail in the interval $[0,t]$ in such a manner as to cause unacceptable damage to other systems or compromise the safety of any people associated with the system.

The safety function can be described by Eq. (2.37):

$$S(t) = P_{\text{functional}}(t) + P_{\text{safe-mode}}(t) \tag{2.37}$$

Safety is directly dependent on "risk", as the probability of loss associated to a particular failure. In turn, risk is a function of the probability of failures and their severity on the system. A system can be unreliable, have low availability and yet be safe. A system is safe if it functions correctly or if in case of failure it can remain in a safe state.

2.5 Security

Security is the concurrent existence of three attributes: integrity, maintainability and availability.

2.5.1 Integrity

The attribute of Integrity is inward-looking and is related to the capability of a system to protect computational resources and data under severe circumstances. Integrity can be defined as the absence of improper system state alterations. As suggested by Storey (1996) two types of integrity can be defined:

- System integrity: the ability of a system to detect faults during operation and to inform to a human operator.
- Data integrity: the ability of a system to prevent damage in data and possibly to correct errors that occur as a consequence of faults.

2.5.2 Maintainability

Qualitatively, maintainability is referred to as the ease and rapidity in which, following a failure, a repairable system can be restored to a specified operational condition. Quantitatively, maintainability is defined as the probability $M(t)$ that a failed system will restore to a normal operable state specified within a given time frame t.

The restoration process involves the location of the problem, the reparation \recovery of the system bringing it back to a normal operational condition. Maintainability has two main components, serviceability and recoverability, that should be carefully analysed in the implementation of self-repairing systems (Eq. (2.38)):

$$M(t) = f(S(t), \mathrm{RC}(t)) \tag{2.38}$$

Maintainability characteristics are determined by the system design of maintenance procedures, such as preventive (PM) and corrective maintenance (CM) procedures that respectively apply to the serviceability and recoverability components and determine the length of repair times (Bodsberg and Hokstad 1995; Dhillon 2006).

PM is the set of activities performed on a system before the occurrence of failure in order to prevent any degradation in its operating condition. PM aims to reduce the probability of failure at predetermined intervals or along with prescribed criteria. CM is the remedial set of activities performed on a system in order to recover an item to its fully functional condition. CM is usually unplanned that requires urgent attention.

Figure 2.13 shows PM and CM mechanisms on a three-state repairable system. Note that not all maintenance leads to downtime of the three-state system. Whilst running PM and remedial CM prevent and correct failures during normal operation, shutdown PC and CM take place during non-functional states.

Figure 2.13 shows PM and CM mechanisms on a three-state repairable system. Note that not all maintenance leads to downtime of the three > state system. Whilst running PM and remedial CM prevent and correct failures during normal operation, shutdown PC and CM take place during non-functional states.

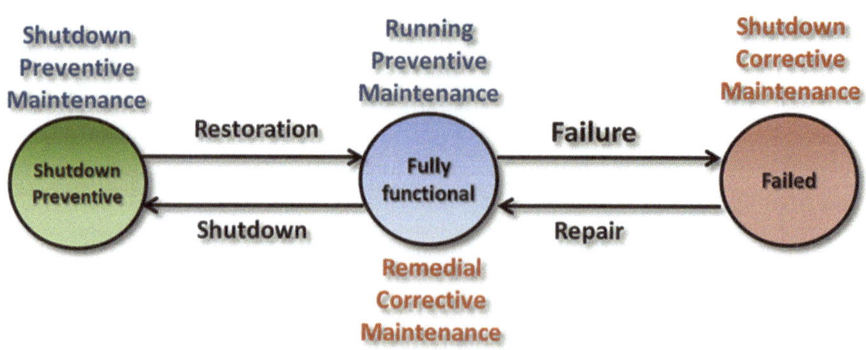

Fig. 2.13 Preventive and corrective maintenance on a three-state repairable system

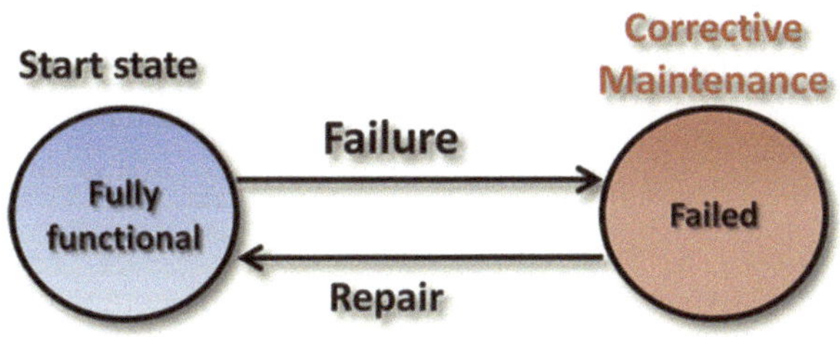

Fig. 2.14 A repairable system with two states and corrective maintenance

2.5.2.1 Recoverability

Once the problem has been identified and located by the testability mechanisms, CM can be carried out to complete the necessary repairs. Consider a repairable system with two states: a fully functional and a failed one (as in Fig. 2.3); however, in this case the failed state can be abandoned after successful CM, transiting back to a fully functional state (as in Fig. 2.14).

Recoverability RC(t) may be defined as the ease of restoring the service after failure. It can be modeled as Eq. (2.39):

$$\mathrm{RC}(t) = 1 - \mathrm{e}^{-\mu t} \tag{2.39}$$

where μ is the repair rate or average number of repairs that can be performed per time unit, the key aspects of recoverability, MTTR and MTTD, are given by Eq. (2.40):

$$\mathrm{MTTD} + \mathrm{MTTR} = \frac{1}{\mu} \tag{2.40}$$

Note that good testability would affect recoverability to a degree. MTTR is further discussed below in the availability section.

2.5.2.2 Serviceability and Testability, $T(t)$

Testability $T(t)$ is the ease in which servicing and inspections can be conducted in order to identify the characteristics of a system; it is the ability to check certain attributes within a system. Measures of testability allow the system to assess the ease of performing tests. Ideally, in order to improve testability the tests can be automated and implemented as an integral part of the system.

These techniques can be used for error detection and error correction within the system. Since most of the time, testability is often used to determine the source of the problem, one way to improve the maintainability of the system significantly is the use of automatic diagnosis.

Testability relates to reliability since it allows detection and correction of errors that would, otherwise become failures, thus improving the overall reliability of the system. Testability is clearly connected with recoverability due to the importance of minimising the time to locate and identify specific problems.

Two properties/measures closely associated to testability, controllability and observability (Franklin and Saluja 1995, p. 199; Goldstein 1979). Observability relates to the probability of "observing", via output measurements, the state of a system.

Controllability instead is associated to the ease of forcing parts of the system into desired states by using appropriate control signals. Design for testability techniques (DFT) (Alanen and Ungar 2011; Karimi and Lombardi 2002; Landis 1989; Mathew and Saab 1993), can be used in order to increase observability and controllability of systems.

2.5.2.3 Coverage

Mathematically, fault coverage C is the conditional probability that, given de existence of a fault in the operational system, the system is able to recover, and continue information processing with no permanent loss of essential information (Bouricius et al. 1969), i.e. Eq. (2.41):

$$C = \Pr\left[\text{system recovers} \mid \text{system fails}\right] \tag{2.41}$$

Fault coverage is a good measure of maintainability and, specifically of the system's ability to detect, locate, diagnose, contain and recover from the presence of a fault. Several types of fault coverage can be distinguished, depending on whether the designer is concerned with fault detection, diagnosis, containment or recovery (Kaufman and Johnson 2001).

In Fig. 2.15, we extend the phases of fault handling by (Dugan and Trivedi 1989), showing the relationship among the steps of recovery and their coverage.

Fault detection coverage C_d measures the system's ability to detect fault. *Fault diagnosis coverage*[4] $4C_l$ is a measure of the system's ability to locate and determine the type of fault. *Fault containment/isolation coverage* C_c is a measure of the

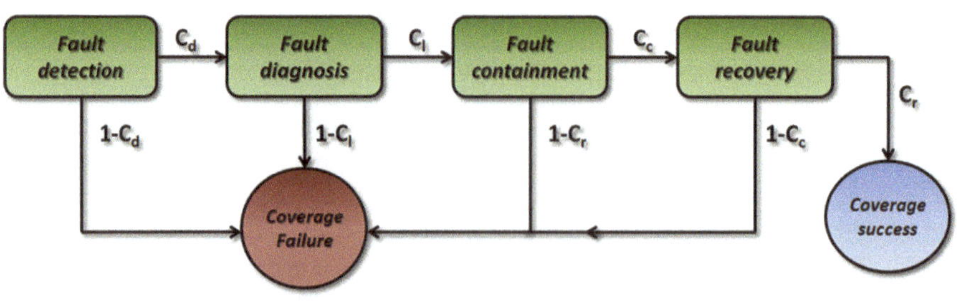

Fig. 2.15 Four phases of fault handling and their coverage

[4] Fault diagnosis involves both the location (*fault location*) and determination of the fault type (*fault determination*).

system's ability to contain faults within a predefined boundary (*fault containment region* or *FCR*). For instance, fault that occurs in a subsystem can be detected, located, and its effects can be prevented from propagating to other subsystems.

Finally, the general term "*coverage*" or "*fault coverage*" is often used to refer *to fault recovery coverage*, which measures the system's ability to *recover from faults* and *maintain correct operation. Recovery* may involve modifying the structure to remove the faulty component (*reconfiguration*) including *graceful degradation. The fault coverage C* for the system is given by Eq. (2.42):

$$C = C_{\mathrm{d}} \times C_{\mathrm{l}} \times C_{\mathrm{c}} \times C_{\mathrm{r}} \tag{2.42}$$

Clearly, high fault recovery coverage requires high fault detection, diagnosis and containment coverage.

2.5.3 *Availability*

A simple definition for availability of a repairable[5] system is "Readiness for correct service" (Avizienis et al. 2004). This measure is suitable for applications in which continuous performance is not essential but where it would be costly to have long downtimes. Availability is strongly dependent on how frequently the system becomes non-operational (reliability) and how quickly it can be repaired (maintainability) (see Fig. 2.14).

As defined in the MTBF Eq. (2.23) the mean time between failures of a system can be defined as a combination of MTTF, MTTR and MTTD. Figure 2.16 illustrates the variations of the state (functional-failed) of a repairable system. The time of operation of such systems is discontinuous. From time 0 to time x_1 the system is continuously available and therefore has an internal availability of 1.

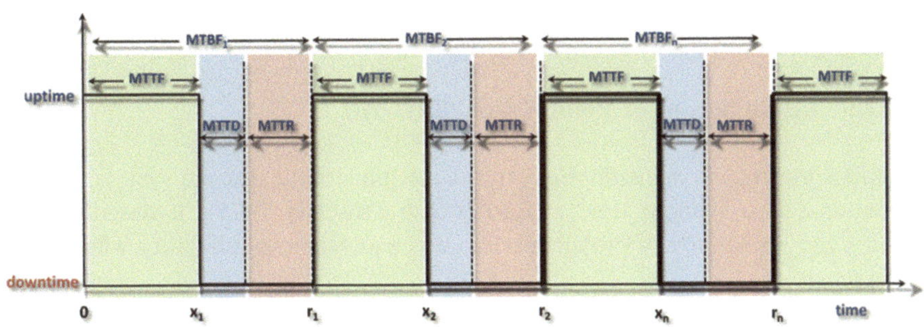

Fig. 2.16 Failure and repair cycle of a system

[5] The concept of availability is applicable to repairable systems. Availability of a non-repairable system would be the equivalent to reliability.

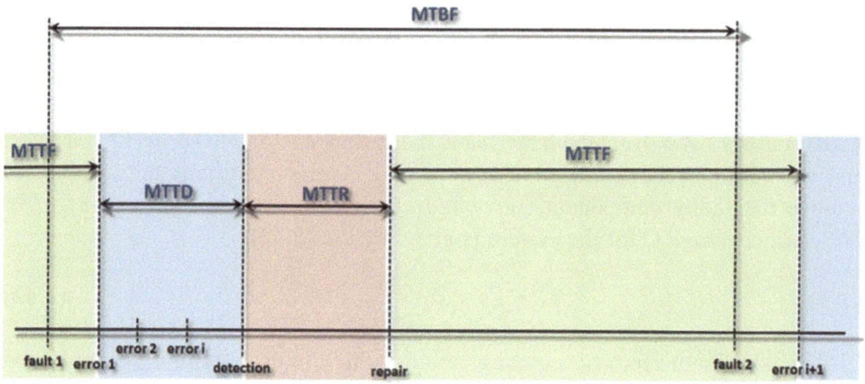

Fig. 2.17 Relation between time to failure (TTF), time between failures (TBF) and time to repair (TTR)

After the first failure at time x_1 internal availability keeps decreasing until the detection and recovery mechanisms complete the repair at time r_1, returning to the original functional state. The system will fail again at time x_2 after a certain time of operation $[r_1 - x_2]$, get repaired at time r_2, and this process will reiterate. Assuming that X_i is an average of system failure and i an average of system repair, for $i > 1$ (Eq. (2.43)):

$$\mathrm{MTBF} = \sum\nolimits_{k=1}^{n} (X_i - X(i-1)) \tag{2.43}$$

The relation between time to failure, time between failures and time to repair is displayed in Fig. 2.17 below.

There are various availability measures that can be classified differently depending on the time interval preferred or the downtimes used.

2.5.3.1 Instantaneous or Point Availability, $A(t)$

Instantaneous or *point availability* $A(t)$ is the probability that the system will be operational at a random time t (Barlow and Proschan 1975). It describes the *on-demand probability* of proper service. It is equivalent to reliability when there is no repair.

While *internal availability* is based on an interval time, instantaneous availability is based on a specific instant of time. At any given time t, the system will be functional if one of the following conditions is met (Elsayed 1996):

- The system was functional from 0 to t (it never failed by time t). The probability of this happening is $R(t)$ (Eq. (2.14)).

- The system has been functional since the last repair time r_i (see Fig. 2.16) when $0 < r_i < t$. This has a probability of functioning since last repair for $0 < r_i < t$ (Eq. (2.44)):

$$\int_0^t R(t - r_i)m(r_i)\mathrm{d}r_i \tag{2.44}$$

With $m(r_i)$ being the renewal density function of the system.

The instantaneous availability of the system is the summation of these two probabilities (Eq. (2.45)):

$$A(t) = R(t) + \int_0^t R(t - r_i)m(r_i)\mathrm{d}r \tag{2.45}$$

2.5.3.2 Average Uptime Availability (or Mean Availability), $\overline{A(t)}$

The average uptime availability or mean availability $\overline{A(t)}$ (Lie et al. 1977) is the proportion of time during a time period $[0 - t]$ that the system is functional and is given by Eq. (2.46):

$$\overline{A(t)} = \frac{1}{t}\int_0^t A(r_i)\mathrm{d}r \tag{2.46}$$

This type of measure is suitable to systems with periodical downtime for maintenance/repairing.

2.5.3.3 Limiting or Steady-State Availability, $A(\infty)$

The limiting or steady state availability (Applebaum 1965) of the system $A(\infty)$ is the limit of the availability function as time t tends to infinity (Eq. (2.47)):

$$A(\infty) = \lim_{t\to\infty} A(t) \tag{2.47}$$

2.5.3.4 Inherent Availability, A_I

In its simplest form, availability A can be mathematically generalised as a function of Up and Down time (Eq. (2.48)):

$$A = \frac{\text{Uptime}}{\text{Uptime} + \text{Downtime}} \tag{2.48}$$

During the design phase of a FT system, Inherent availability A_I is a useful measure (Valstar 1965). A_I defines the availability of a system only in regard to effective functional time (uptime) and downtime due to corrective maintenance (CM). It can be calculated using estimated parameters (MTTF, MTTD and MTTR) as Eq. (2.49):

$$A_I = \frac{\text{MTTF}}{\text{MTTF} + \text{MTTD} + \text{MTTR}} = \frac{\text{MTTF}}{\text{MTBF}} \tag{2.49}$$

Hence, if MTTF or MTBF are long compared to MTTR and MTTD, then the system's availability will be high. Likewise, if MTTR and MTTD are short, then the system's availability will also be high. As reliability decreases (e.g. low MTTF), better recoverability will be needed (lower MTTR/MTTD) to achieve the same availability.

2.5.3.5 Achieved Availability, A_{A}

A_{I} is a good parameter to measure systems under ideal conditions where downtime due to preventive maintenance (PM) is overlooked. Achieved availability A_{A} is similar to inherent availability with the exception that downtimes due to PM tasks are also included (Conlon et al. 1982). According to US department of defense the achieved availability is defined by Eq. (2.50):

$$A_{\text{A}} = \frac{\text{OT}}{\text{OT} + \text{TCM} + \text{TPM}} \tag{2.50}$$

Where OT is the total operating time, TCM is the total corrective maintenance time and TPM is the total time spent during preventive maintenance actions.

2.5.3.6 Availability–Recoverability–Reliability Relationship

At first glance, it might seem that a highly available system would also have high reliability. Nonetheless, this in not always the case, a system can be highly available yet suffer from frequent periods of non-operation as long as the length of the downtime is extremely short. Let us explore further the relationship between availability and reliability.

Reliability represents the probability of systems and components to perform its intended function for a desired period of time [0, t] under specified environmental and operating conditions. However, reliability in itself does not take into account any repair actions. Reliability does not reflect how long the recovery of a component/system will be needed in order to take it back to a working condition.

Availability reflects not only how often a system fails but how often it can be repaired (it accounts for repair actions). Thus, it is a function of reliability, recoverability and testability (Eq. (2.51)).

Table 2.1 Reliability–recoverability–availability relationship

Reliability	Recoverability	Availability
Constant	Constant	Constant
Constant	Decreases	Decreases
Constant	Increases	Increases
Decreases	Constant	Decreases
Increases	Constant	Increases

$$A(t) = f(R(t), M(t)) \tag{2.51}$$

Table 2.1 below presents the relationship between reliability, recoverability and availability. As shown by the table, once again, high reliability does not necessary implies high availability. Availability decreases as time to repair increases. Even an unreliable system could present high availability if MTTR is low.

2.6 Performability

The all-or-none nature of operation implicit in classic reliability and availability models does not measure in detail systems that can operate with different capability levels (e.g. multiprocessor systems). Consequently, another key attribute of resilience, performability and its measure, mean computation before failure (MCBF) can be employed. MCBF is described as the expected amount of computation available on the system before its first failure, given an initial state (Beaudry 1978).

In qualitative terms, performability is the ability of a system or component to accomplish its designated functions within specified constraints such as speed, accuracy or memory usage. It is the measure of the likelihood that some subset of the functions of the system or component is performed correctly during a certain time interval. Quantitatively, Performability $P(L,t)$ has been defined as "the probability that the system's performance will be at or above some level L at the instant of time t" (Fortes and Raghavendra 1985).

After the occurrence of faults and errors, certain systems have the ability to continue to perform correctly, however with a diminished level of performance. This ability or feature is called Graceful degradation, or fail soft operation (Gountanis and Viss 1966), and it is the ability of a system (gracefully degrading system or GDS), upon failure of one or more of its component units, to continue the processing of tasks at the expense of decreasing its performance level. The performability of a GDS $P(L,t)$ at time t depends on the amount of available resources and their computational capability provided.

Note that performability differs from reliability in that reliability measures the likelihood that all functions are performed properly, whereas performability measures the likelihood that some subset of the functions is performed properly.

2.7 Resilience

Historically, the term resilience has had multiple meanings in various fields. As a property it has different connotations. In social psychology resilience is about elasticity, spirit, resource and good mood. On the other hand, in material science resilience involves not only elasticity but robustness.

In computer science it has been identified as a synonym for fault tolerance. In this book we adopt to extend the concept of resilience for safety critical applications. First we start by selecting the material science connotations. Hence, our definition of resilience includes both attributes: robustness and elasticity.

Figure 2.18 illustrates the different attributes and measures of resilience. The term robustness involves the use of static techniques such the use of very reliable materials or the use of rigid and pre-design approaches of fault tolerance. A robust system can deliver correct service in conditions beyond the normal domain of operation without fundamental changes to the original system.

This is more an aim than an objective. Total reliability to unforeseen faults other than the normal domain of operation is not feasible.

On the other hand, we interpret elasticity as the ability to "spring back" (or recover) without losing the intrinsic properties of the material (in our case a system). Applied to resilience, we understand elasticity as the ability to evolve, to successfully accommodate changes (evolvability). An evolvable system may perform changes to the system, decreasing its level of performance or reliability for a specific time range (1) to compensate for faults or (2) during exceptional circumstances (graceful degradation).

More specifically, we consider that a resilient system must have the ability to be adaptable, understanding adaptability as the ability to evolve while executing. Therefore, adaptability is a subset of evolvability and requires the ability to anticipate to changes prior to the occurrence of the resulting damage.

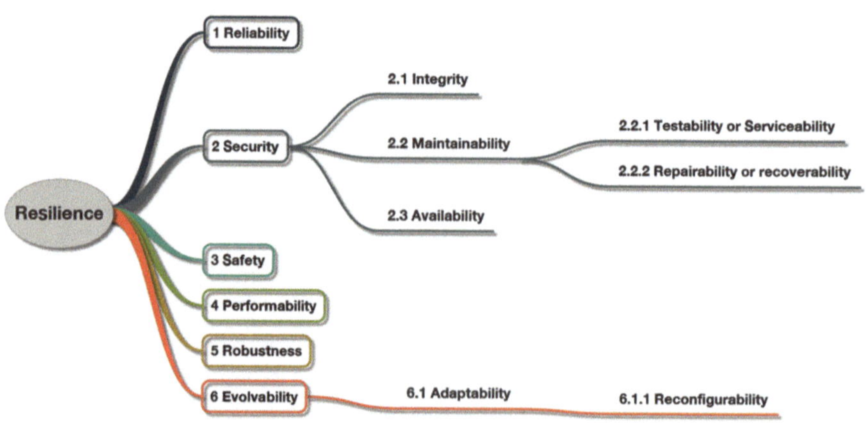

Fig. 2.18 Attributes and measures of resilience

Therefore, a resilient architecture must include different mechanisms to acquire both attributes: (a) static pre-design fault tolerant techniques (robust) and (b) dynamic techniques (elastic) that may be achieved with the ability to reconfigure elements of the system (reconfiguration).

2.7.1 Requirements

The main aim required to implement a resilient architecture for safety critical applications is the "ability to deliver correct service adapting to disturbance, disruption and change within specified time constraints".

The above aim can be subdivided into more specific objectives, as follows:

- Continuity of service (reliability).
- Readiness for usage (availability).
- Non-occurrence of catastrophic consequences (safety).
- Non-occurrence of incorrect system alterations (integrity).
- Ability to undergo corrective maintenance and recovery with maximum coverage of faults (testability and recoverability).
- Ability to perform in the presence of faults (performability).
- Ability to decrease the level of performance for a specific time range in order to compensate for hardware faults (graceful degradation).
- Ability to regain operational status via reconfiguration in the presence of faults (recoverability via reconfiguration).
- Ability to accommodate changes (evolvability).
- Ability to anticipate to changes (adaptability).

2.7.2 Effectiveness of Resilience

Being reliability $R(t)$, security $\mathrm{SC}(t)$, integrity $I(t)$, maintainability $M(t)$, testability $T(t)$, recoverability $\mathrm{RC}(t)$, availability $A(t)$, safety $S(t)$, performability $P(L,t)$, robustness $\mathrm{RB}(t)$, evolvability $E(t)$, adaptability $\mathrm{AD}(t)$ and reconfigurability $\mathrm{RC}(t)$. Maintainability is a function of serviceability and repairability (Eq. (2.52)):

$$M(t) = f(T(t), \mathrm{RP}(t)) \tag{2.52}$$

Security is a function of integrity, availability and maintainability (Eq. (2.53)):

$$\mathrm{SC}(t) = f(I(t), A(t), M(T(t), \mathrm{RP}(t))) \tag{2.53}$$

Evolvability is a function of adaptability and reconfigurability (Eq. (2.54)):

$$E(t) = f(\mathrm{AD}(t), \mathrm{RC}(t)) \tag{2.54}$$

Therefore, resilience RES(t) will be a function of all these attributes (Eq. (2.55)):

$$\mathrm{RES}(t) = f\begin{pmatrix} \text{reliability, integrity, testability, recoverability,} \\ \text{availability, safety, performability,} \\ \text{robustness, adaptability, reconfiguration} \end{pmatrix} \tag{2.55}$$

With all these attributes the following systems would benefit from the implementation of effective resilience:

- Safety-life critical: e.g. aircraft and nuclear reactor control, life support systems.
- Business critical.
- Reliability-critical: e.g. telephone switching-, traffic light control-, automotive control (ABS, fuel injection) systems.
- Mission-critical and long life systems: e.g. manned and unmanned space-borne satellites and other systems in inaccessible locations.
- High availability such as transaction processing and non-stop systems.

Resilience is not a simple and single concept, rather, it possesses different components or key attributes. Taking into consideration all these attributes, our definition of resilience is as follows:

> A resilient system is a system that over a specified time interval, under specified environmental and operating conditions, is ready to perform its intended function, guaranteeing the absence of improper system alterations, having the ability to anticipate and accommodate changes while executing, and the ability to conduct servicing and inspections so that in case of failure quick restoration to a specified working condition must be achieved, or otherwise discontinue of the operation in a safe way is provided.

2.8 Conclusion

This chapter explains the concept of resilience that encompasses important attributes and measures that are used further. Such concepts have been introduced before attempting to explain resilience itself.

Safety critical systems must provide correct service at all times by trying to avoid the occurrence of any catastrophic failure. Different techniques can be employed to increase reliability by avoiding/preventing hardware faults from becoming errors that may lead to failures and catastrophic failures.

We introduce the concept of vicious cycle that explains the reasons behind the performance and reliability problems that the microprocessor industry is currently facing. The increase of transistor density, operating frequencies and architectural complexity is drastically decreasing the reliability of newer systems. There is, therefore, a need for implementing mechanisms that can deal with the upcoming fault rates.

The essential background for classical reliability has been presented. The basic definitions for reliability evaluation have been reviewed. It has been shown that for constant failure rate, independent of time, the exponential distribution is the most suitable for the reliability analysis of the useful time of systems. The age of a system should be taken into account when analysing reliability. Three different periods of system life with different reliability have been explained by introducing the bathtub curve, which represents the effect that ageing and degradation have on HW reliability.

We have briefly described an estimation of reliability for serial, parallel and mixed systems. The failure rate of a serial system is equal to the sum of the failure rates of its individual elements. Therefore, the more components a serial system has, the higher the probability of system failure. The reliability of a system is often dominated by the reliability of its least reliable component. Introducing deliberately and carefully extra components into a system, overall reliability can be increased, provided reliability benefit accruing from the redundancy scheme exceeds the decrease in reliability due to the actual implementation of the redundancy mechanisms itself.

We extend the classical definition of resilience and apply it to the field of safety critical computing. Moreover, we quantify the key attributes that a resilient system must have, exploring the relationships among these quantitative measures. The attributes of safety and performability are explained. The concept of security is described including attributes: integrity, availability, testability and recoverability. The mathematical background and the basic definitions for system availability are also introduced.

We show how the availability of FT systems can be estimated using different methods and measures.

Finally, the pathways of design and development of resilient architecture for safety critical applications are defined. Property-wise descriptions of resilient systems are explained.

It is pointed out that a resilient system, over a specified time interval, under specified environmental and operating conditions (performability), "must be ready" (in terms of availability) to perform its intended function (with reliability), guaranteeing the absence of improper system alterations (integrity).

A resilient system must have the ability to conduct servicing and inspections (testability) so that in case of failure achieving quick restoration to a specified working condition (maintainability) can be provided or it can discontinue its operation in a safe way (safety). Furthermore, a resilient system must have the ability to anticipate changes and evolve (evolvability) while executing (adaptability), successfully accommodating changes by reconfiguring elements of the system if necessary (reconfiguration).

Chapter 3
Dealing with Faults: Redundancy

3.1 Handling Faults: Design Strategies

In order to increase the reliability of safety-critical systems so that correct service can be delivered within predefined "envelope of requirements", we need to develop methods and techniques enables us to prevent or reduce the appearance of faults that could cause catastrophic failures. Depending on the phase of the development cycle and the level of abstraction at which the faults are tackled, two different design strategies can be adopted: *fault avoidance* and *fault tolerance*.

Fault avoidance strategies can be applied at device level during design time. Typical in mainstream applications, in order to reduce the number of failures, this approach focuses on preventing the occurrence of faults. Since a failure is the consequence of an error propagating, and an error is the consequence of a fault, eliminating faults would improve reliability. Examples of fault avoidance are: silicon on insulator (SOI) and hardened memory cells. These techniques and supportive technologies have drawbacks in terms of cost, speed of operation, chip area and power consumption.

In turn, at execution or, often called "run-time", *fault tolerance* should be implemented. Fault and tolerance to it can be considered at different levels of abstraction, fault tolerance strategies can be implemented at a system level or element level.

Following the failure life cycle and its different phases already described in Sect. 2.1, Fig. 3.1 adds the different mechanisms to deal with faults within the fault generation, error activation and failure propagation phases. Additionally, Fig. 3.1 serves as a summary of the chapter introducing the fault avoidance and fault tolerance techniques and their phase of interaction within the failure life cycle.

V. Castano, I. Schagaev, *Resilient Computer System Design*,
DOI 10.1007/978-3-319-15069-7_3

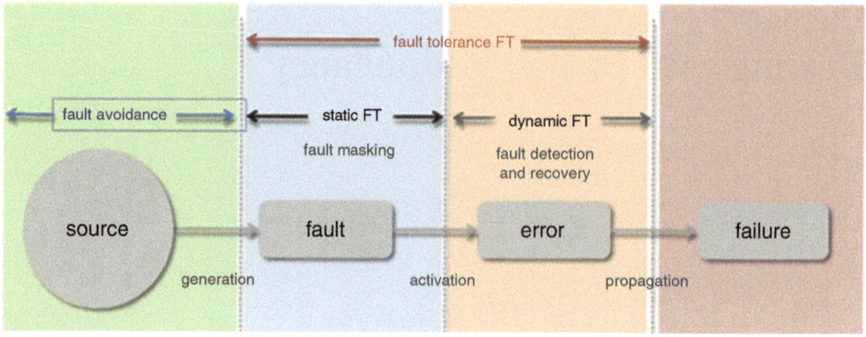

Fig. 3.1 Mechanisms to deal with faults within the fault-failure life cycle

Focusing on the source of faults, fault avoidance mechanisms attempt to prevent faults from occurring in the first place. Once a fault has been generated it can be prevented from activating an error using static fault tolerant techniques such as masking. Alternately, errors can be detected and recovered using dynamic fault tolerance techniques. Therefore, either we prevent the faults from taking place (fault avoidance) or we deal with them using fault tolerance techniques.

3.2 Fault Avoidance

Nowadays, mainstream systems employ fault avoidance design strategies in order to achieve their projected failure rates. Manufacturing companies perform assessments of sources and weaknesses that could lead to potential failures. Based on the assessments, preventive measures are taken to ensure that the overall reliability target is not compromised.

Additionally, fault avoidance strategies may include technology and design mitigation techniques that implicate modifications of conventional manufacturing processes. These techniques involve the use of specific materials, the modification of the doping profiles of devices and substrates and the optimisation of deposition processes for insulators.

Technology mitigation techniques consist of IC process variations by either improving the manufacturing process or by improving the materials used.

Improving materials: implicates the selection of specific materials with better characteristics. For example:

– Boron has been used extensively as a p-type dopant in silicon and has also been used in Boron Phosphorus Silicate Glass (BPSG) dielectric layers. For BPSG-based semiconductor processes, BPSG can, in fact, be the predominant source of transient errors (Baumann 2001). The removal of B-10 Boron isotopes in BPSG has been proven effective in the reduction of transient errors (Baumann et al. 1995).

- *Lead-free* materials can reduce the effect of alpha particles (May 1979) (extensive information on alpha particle effects is provided in Chap. 4).
- Implanting of elements such as A_l, A_s, F_l, P and S_i into oxides improves the resilience to Total Ionising dose effects (TID) (Kato et al. 1989; Mrstik et al. 2000; Nishioka et al. 1989).

I*mproving the manufacturing process* is based on changing the charge collection and charge sharing capabilities of the devices:

- *Substrate + techniques,* e.g. using epitaxial substrate doping (EPI layer charge reduction)(Puchner et al. 2006), wells (single well, twin well and triple well processes) (Pellish et al. 2006; Puchner et al. 2006; Roche and Gasiot 2005), buried layers (Roche and Gasiot 2005) and dry thermal oxidation (Hughes and Benedetto 2003)
- *Non-capacitance techniques*, e.g. increasing the node coupling capacitance between storage nodes and memory, or using a DRAM capacitor on top of the memory cell (Geppert 2004)
- Using *alternative insulating substrates*, e.g. the use of Silicon on Insulator (SOI) or Silicon on Sapphire (SOS) (Schwank and Dodd 2003) would mitigate significantly the transient faults due to radiation (described in Chap. 4).

Whilst technology mitigation techniques are based at the process level, **design mitigation techniques** operate at the layout level. An example of this type of technique is the use of enclosed layout transistors. Furthermore, to prevent the effects of radiation memory cells can be hardened with the use of contact and guard rings.

The effect of silicon failure mechanisms, such as radiation induced transient faults and wear-out defects, is proportional to the clock speed, supply voltage, temperature, etc. Therefore, to ensure system reliability safety margins are inserted into clock speed, operating temperature and supply voltage margins. If the system failure rates resulting from the use of fault avoidance strategies fall within the specified reliability targets, the use of redundancy techniques is not justified. However, this is not the case for safety-critical systems.

Despite all the testing, verification techniques and technology improvement, hardware components will eventually fail. The fault avoidance approach will not be panacea and will be insufficient if:

- Failure rate and MTTR are unacceptable or,
- The system is inaccessible for repair and maintenance actions

Therefore, fault avoidance techniques are only part of the solution for real time safety critical domains. Complete removal of faults via fault avoidance is not possible; above all, it has drawbacks in terms of cost of manufacturing the elements required, speed of operation and increased chip area.

3.3 Fault Tolerance: Using Redundancy

The key ingredient of fault tolerance is redundancy. Redundancy is defined as the addition of information, resources or time beyond what is needed for correct system operation (Latchoumy et al. 2011). Fault tolerant techniques rely on redundancy that may include a combination of additional elements of hardware and/or software to detect and/or recover from faults. These components are called redundant since they are not required in a *perfect + system.*[1]

Artificially built-in or *protective redundancy* is a system property that we define as the incorporation of extra components (transistors at a low level) in the design of a system so that its function is not impaired in the event of a failure. Redundancy may arise by design (artificially built-in redundancy) or as a natural by-product of design (*natural redundancy*). Natural redundancy is usually unexploited whilst artificially built-in redundancy has been deliberately introduced. In this work, when the term redundancy (or redundant) is used it is meant to have the artificial connotation instead of the natural one.

When a system does not provide the minimum reliability required, extra redundancy, not strictly necessary for the normal functioning of the system, can be added in order to increase the probability of normal functioning. Notice that that the term redundant does not mean identical functionality; it just denotes that it performs the same task. In this sense, heterogeneous hardware performing the same work can also provide redundancy.

Fault tolerance assumes actions such as fault detection, location of the faulty component, recovery and, if necessary, reconfiguration of the system. Fault detection is the process of determining the presence of faults and the time of occurrence. Fault location is to exactly locate the reason/origin of the fault. The system must be dynamically restored as though it is "*as good as new*" in operational terms, except for the fact that some of the redundancy has been used up and this may limit the possibilities for future repairs.

Several attempts to classify redundancy have been made (Avizienis 1971; Carter and Bouricius 1971; Schagaev 1989, 2001). This work follows the approach proposed by (Schagaev 2001).

Figure 3.2 shows the different types of redundancy (at the top of the figure) and the way it can be implemented (at the bottom of the figure). In general, three types of redundancy exist: structural (S), involving multiplication of components, information (I), involving multiplication of information/data and time redundancy (T), involving multiplication of functions in time. These can be implemented in hardware and software. We mostly focus on the hardware aspect of redundancy and fault tolerance.

[1] A perfect system is a system with a theoretical 100 % reliability. A perfect system is usually assumed to model extra reliable systems.

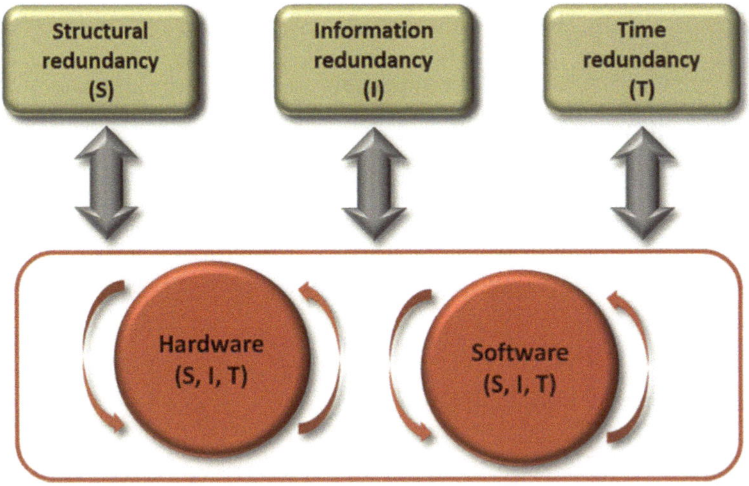

Fig. 3.2 Redundancy types and their implementation

Redundancy comes with a cost. Information and structural redundancies require additional hardware components, extra power and perhaps extra area and shielding. Time redundancy requires faster processing to achieve the same performance, which in turn requires extra hardware and power. Software redundancy involves higher development and maintenance costs.

3.3.1 Redundancy Notation

Existing implementations of system redundancy use at least one of these three redundancy types, usually more than one and can be implemented in hardware HW (), software SW() or a combination of both (HW(), SW()). As an example, hardware based information redundancy is abbreviated as HW(I). Additional quantifiers are used together with the redundancy type to further specify the used redundancy as shown below in Table 3.1:

Note that the current notation does not include the implementation level. HW ($2S$) only indicates duplication, but not whether the whole system is duplicated or it is just parts of that system, such as, for example, duplicated memory.

Tables 3.2 and 3.3 present some concrete examples of notation of hardware and software based redundancy. Any type of redundancy (hardware and software) needs additional structural redundancy for its implementation.

For instance, instruction repetition HW(nT) needs additional hardware registers to store the internal state to be able to perform instruction rollback. We refer to this as *supportive redundancy* and we define it as the redundancy needed for the implementation of the main redundancy technique. For the sake of simplicity, we usually omit this supportive redundancy.

Table 3.1 Redundancy classifiers

Quantifier	Example	Description
	SW(I)	No quantifier means general, not further specified redundancy. SW (I) for instance just indicates general software information redundancy
δ	SW(δI)	Additional used software based redundancy
Number	HW($2S$)	The number indicates duplication (2), triplication (3), etc. of a system if used as a prefix for the redundancy type. The original system and the copies are identical. n instead of a discreet number is used to mark repetition until success in case of time redundancy
Indices	HW (S_1, S_2)	Indices are used to mark a duplicated system implementation/hardware components

Table 3.2 Examples of notation of HW based redudancy

Redundancy type	Description
HW($2S$)	Structural (material) redundancy of hardware such as duplicated memory system
HW(S_1, S_2)	A duplicated FT computer system with principally non identical parts
HW(δI)	Redundant information bit, for example an additional parity bit per data word in HW memory for error detection
HW(nT)	Special HW to delay execution (like in a timing diagram) to avoid transient faults
HW(δT)	Special HW to delay execution to avoid transient faults

Table 3.3 Examples of notation of SW based redundancy

Redundancy type	Description
SW($2S$)	Structural (material) redundancy of hardware such as duplicated memory system
SW(S_1, S_2)	A duplicated FT computer system with principally non identical parts
SW(δI)	Redundant information bit, for example an additional parity bit per data word in HW memory for error detection

In some cases an applied redundancy of several types can be used. An example of this is the case of software based implementation of hardware checks (or tests) that are performed during idle time of the system: SW(δS,δT).

3.4 Structural Redundancy HW(S)

Structural hardware redundancy involves multiple independent hardware components and assumes execution of the same computation over such components at the same time. When this redundancy type is applied for reliability purposes the errors are exposed by checking/comparing the results of the independent executions.

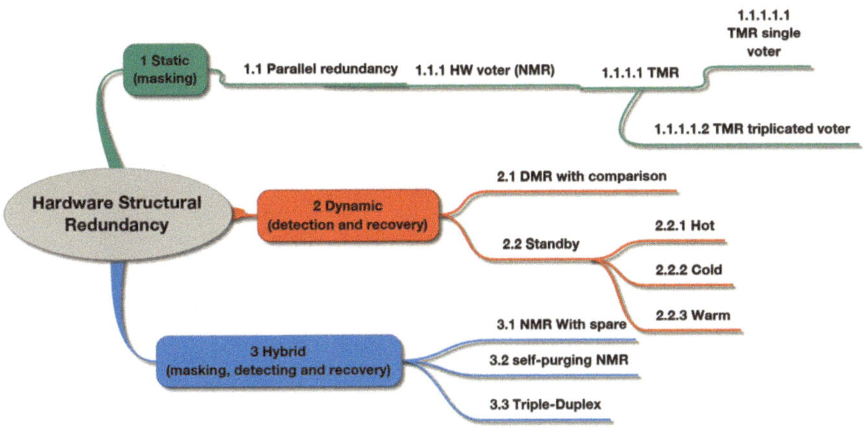

Fig. 3.3 Taxonomy of structural HW redundancy

In terms of granularity, redundancy in general, and not only hardware redundancy, can be applied at different abstraction levels of system description. From bottom up we can distinguish between finer-grained: transistor level, gate or logic level, and between coarser-grained designs: circuit level, function level, system level, microcode level and chip level of abstraction. Therefore, redundant components can be as simple as transistors or logic gates but also as complex as processors or even larger entities: boards, clusters, network segments.

Figure 3.3 illustrates a taxonomy of the different hardware techniques based on structural redundancy. Two different organisations of structural redundancy can be distinguished:

Parallel redundancy with redundant components running concurrently and *Standby redundancy* with a spare component being activated upon failure of an active component.

Furthermore, these extra resources can be used passively (*passive redundancy*), actively (*active redundancy*) or combined. In systems with active redundancy, all redundant components are in operation, sharing the load with the normal components. This implies that both, regular and redundant components, age together. Passive components are not fully energised and start normal operation only when normal components fail. Passive components can be further broken down into two types: *warm* and *cold standby*.

Warm standby components remain partially energised until becoming active and tend to deteriorate with time, hence, having lower failure rate than the regular components. *Cold standby* components are kept in reserve and they only become energised when put into use. These types of components have a zero failure rate, meaning they do not fail when they are in standby mode.

Whilst passive components are switched off completely, standby components are partially activated. Standby redundancy is usually applied when the start time of the component is unacceptably long.

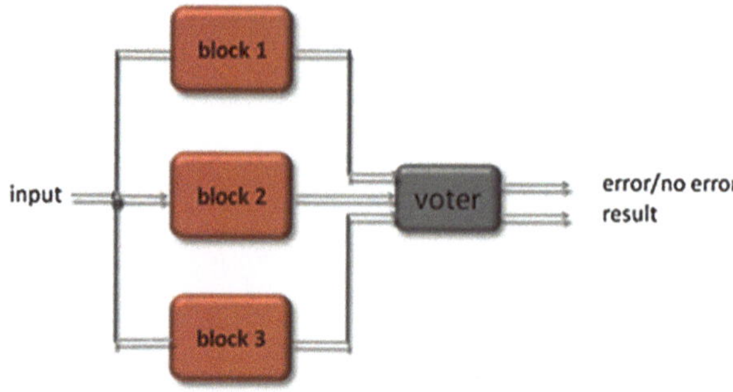

Fig. 3.4 Triple modular redundancy (TMR) with a voter

3.4.1 Static Redundancy

Static redundancy, also called masking redundancy, implements error mitigation. The term static relates to the fact that redundancy is built into the system structure. Fault tolerant techniques based on this type of redundancy (*static fault tolerance*) transparently remove errors on detection. The most common form of hardware redundancy is *Triple modular redundancy* (*TMR*) (Von Neumann 1956) and its generalisation N-modular redundancy (NMR). Note that *Dual modular redundancy* (*DMR*) (DMR is further explained in Sect. 3.4.2.1) is not considered static redundancy since the mismatch can take place but recovery is not possible.

3.4.1.1 Triple Modular Redundancy: HW(3S) + HW(δS)

A basic TMR system (*two-out-of-three*) is a fault tolerant form of NMR that consists of three fully redundant and active components or modules working in parallel with equivalent functionality (Johnson 1989; von Neumann 1956). Figure 3.4 below presents an example of a TMR system with a voter.

The three components perform a process based on individual inputs whose results are in turn processed by a voting system to produce a single output. The voting is based on majority; if any of the three components has a fault, the other two systems can mask the fault. It is assumed that two out of three modules must deliver correct results. Therefore, TMR is capable of masking a single error. Generally, a majority voting mechanism should:

- Guarantee a majority vote on the input data to the voter
- Determine the faulty block

In order to guarantee the majority vote, loosely synchronised systems require synchronisation of the inputs to the voter.

A specific example of this technique is the *Boeing TMR 777* primary flight computer (Yen 1996), which has triple redundancy for all hardware including computing system, communication paths, electrical and hydraulic power.

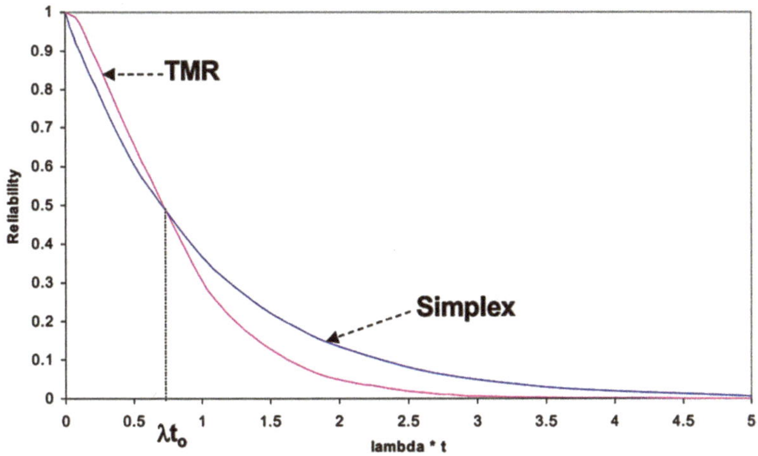

Fig. 3.5 Reliability of TMR vs. simplex system

3.4.1.2 Comparing the Reliability of Simplex and TMR with Perfect Voter[2] Systems

A simple TMR system such the one in Figure 3.5 includes three blocks, two of which are required for the system to provide correct service. Given the reliability of a single component or simplex system by Eq. (3.1):

$$R_{\text{simplex}} = \mathrm{e}^{-\lambda t} \tag{3.1}$$

where λ is the failure rate for the single component; the MTTF of a simplex system can be expressed as:

$$\text{MTTF}_{\text{simplex}} = \int \mathrm{e}^{-\lambda t} = \frac{1}{\lambda} \tag{3.2}$$

The reliability of a TMR system with a perfect voter is given by:

$$\begin{aligned} R_{\text{TMR}} &= R_m^3 + \binom{3}{2} R_m^2 (1 - R_m) \\ R_{\text{TMR}} &= \mathrm{e}^{-3\lambda t} + \binom{3}{2} \mathrm{e}^{-2\lambda t}\left(1 - \mathrm{e}^{-\lambda t}\right) \\ R_{\text{TMR}} &= 3\mathrm{e}^{-2\lambda t} \end{aligned} \tag{3.3}$$

[2] A perfect voter is a voter with a theoretical 100 % reliability. A perfect voter is usually assumed to model extra reliable voters.

and, therefore:

$$\mathrm{MTTF}_{\mathrm{TMR}} = \frac{3}{2\lambda} - \frac{2}{3\lambda} = \frac{5}{6\lambda} \tag{3.4}$$

$$\mathrm{MTTF}_{\mathrm{simplex}} > \mathrm{MTTF}_{\mathrm{TMR}}$$

Figure 3.5 shows how TMR has higher reliability than Simplex for short missions ($t < t_0$). Note that:

$$R_{\mathrm{TMR}}(t) \geq R(t) \quad 0 \leq t \leq t_0 \tag{3.5}$$

$$R_{\mathrm{TMR}}(t) \geq R(t) \quad 0 \leq t \leq \infty$$

Where:

$$t_0 = \frac{\ln 2}{\lambda} \approx \frac{0.7}{\lambda}$$

TMR is assumed to be useful in aircraft applications claiming 0.99999 reliability over a 10-h period. Ravishankar and Iyer (2003) shows that TMR is not suitable for long safety-critical missions ($t > t_0$) because paradoxically, after the first failure, the two remaining components compete to fail. Higher reliability can be achieved extending TMR to N-Modular Redundancy. Therefore, a blind use of redundancy can lead to seemingly paradoxical results. Further we show other serious drawbacks of TMR schemes.

Reliability of TMR with Voting

The previous expression of reliability of TMR assumes that the voter is perfect, that is to say that the voter is 100 % reliable.

The reliability of a generic TMR system with non-perfect single voting (TMRV) and identical blocks is given by:

$$R_{\mathrm{TMRV}} = R_V \left(R_m^3 - \binom{3}{2} R_m^3 (1 - R_m) \right) \tag{3.6}$$

Where R_v is the reliability of the voter mechanism and R_m is the reliability of the block. In terms of reliability, the voter becomes the weak part of this configuration. The voter is a *single point of failure* (*SPF*); if the voter fails then the complete system will potentially fail. This can be tackled by the following different alternatives:

- By increasing the reliability of the voter using fault avoidance techniques
- By triplicating the voter and connecting the module outputs to all three voters (Johnson 1989) so that individual voting failures can be corrected by the extra voting process
- By implementing online self-testing for the voting circuitry (Cazeaux et al. 2004; Metra et al. 1997)
- Using an I_{DDQ} *checkable voters* (*ICVs*) (Bogliolo et al. 2000): under fault-free conditions, *ICVs* work as traditional *CMOS* voters; however, they cause *quiescent supply currents* (*IDDQs*)[3] in the presence of maskable stuck-at faults (see Sect. 5.3.1). Faults can be detected using I_{DDQ} *testing*, by monitoring *IDDQs* (Williams et al. 1996)

A basic TMR system does not support common-mode failures (*CMFs*[4]) (Lala and Harper 1994). *CMFs* are the result of failures affecting more than one component, usually due to a common cause, which may be due to design-faults or operational ones resulting from external (such radiation or electro-migration) or internal causes.

For instance, a radiation source causing multiple event upsets (Reed et al. 1997) can potentially lead to the failure of more than one component in a TMR system.

3.4.1.3 N-Modular Redundancy: HW(nS) + HW(δS)

The generalised version of TMR is NMR where *N* stands for the number of redundant modules. The main advantage of using *N* modules as opposed to only three is that often more faults can be tolerated. For instance, a *5MR system* contains 5 replicated modules including a majority voting arrangement. The voter allows the system to deliver correct service in case of as many as two module faults.

Figure 3.6 depicts a generic N-modular redundant system with a voter. The redundancy of this system can be defined as HW(nS) + HW(δS) using the previous notation. NMR works similarly to TMR but this type of structure is able to detect $[(N-1)]/2$ errors in different processing modules. Besides TMR, 5- and 7- modular redundancies are the most common structures and are capable of detecting two and three errors respectively.

M-out-of-*N* systems are a type of NMR. The reliability of a generic *M*-out-of-*N* system assuming that it has a perfect voter and *M* out of *N* modules need to function is expressed by:

[3] Quiescent current is the current consumed by a circuit when no load is present. Fault-free CMOS devices have very quiescent currents when they are in a quiescent state. Faults that cause high quiescent currents can be detected if the quiescent current is significantly higher that that of a fault-free circuit (Williams et al. 1996).

[4] Multiple faults can be either independent (attributed to different causes) or related (attributed to a common cause. Both can lead to similar errors (e.g. errors that cannot be distinguished by the detection mechanisms being used) (Avizienis and Kelly 1984). The failures triggered by similar errors are called CMF.

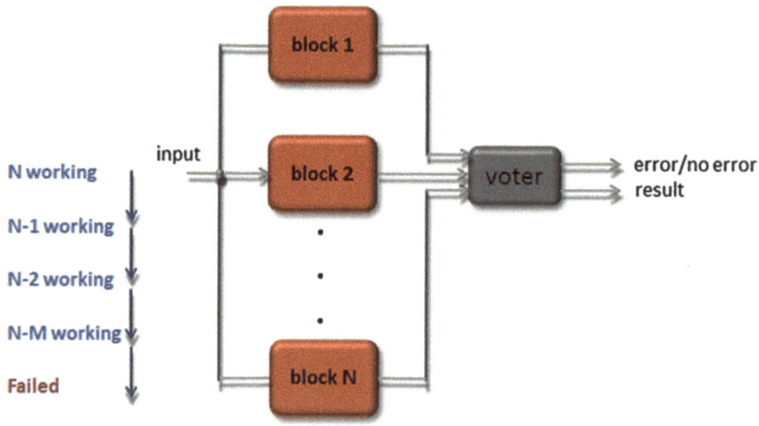

Fig. 3.6 N-modular redundancy with a voter: M-out-of-N system

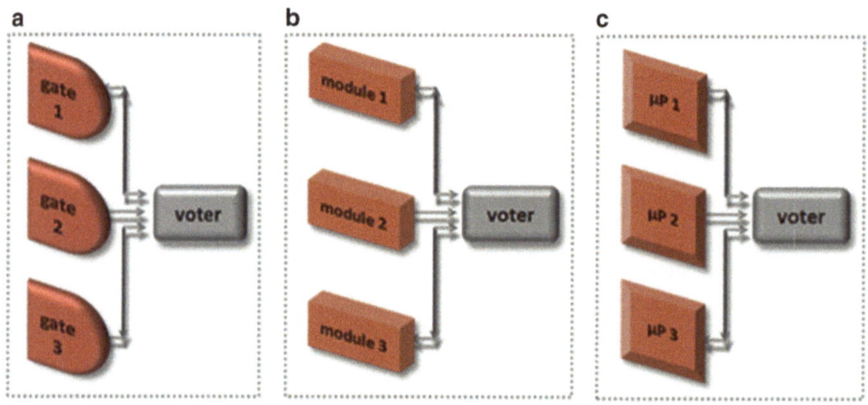

Fig. 3.7 Redundancy applied at different levels of abstraction: (**a**) Three logic gates in a TMR at the logic or gate level of abstraction; (**b**) Three memory modules in a TMR configuration at the circuit abstraction level; (**c**) Three microprocessors in a TMR configuration at the chip level

$$R_{MN} = \sum\nolimits_{i=0}^{N-M} \binom{N}{i} R_m^{N-i} (1 - R_m)^i \tag{3.7}$$

Note than NMR systems offer higher reliability than TMR but at a much higher cost. Undoubtedly, for practical applications there must be some limit on the amount of redundancy that can be employed.

TMR and NMR could be applied at different levels of abstraction triplicating logic gates, single memory cells, memory modules or complete microprocessors. Figure 3.7 displays how TMR can be applied at logic (a), circuit (b) and chip level (c).

TMR and NMR are typically employed in aerospace applications where the cost of failure is particularly high. However, the higher reliability of these systems involves more than 200 % increase in redundancy. Such an example is the *NASA* Space Shuttle on-board system, which is based on four computers with a majority voter (Sklaroff 1976).

3.4.2 Dynamic Redundancy

To reduce the extensive space, energy and performance overheads of TMR and NMR systems, numerous approaches have been developed. These approaches are usually based in dynamic redundancy, which implements error handling. This type of redundancy is similar to static redundancy with the main difference being the voter logic is replaced with a switch that is controlled by an error detection block.

At least one of the modules is working as the main module, whereas the rest of the modules or replicas can either be working in parallel (e.g. DMR with comparison) or can be turned off and used as spares (stand-by redundancy).

To avoid failures, after a fault has been detected, the system must be reconfigured. Detection, reconfiguration and recovery are required in order to prevent error propagation. Some examples of this type of redundancy are: pair and spare, duplex systems (DMR with comparison), backup sparing techniques etc.

3.4.2.1 Dual Modular Redundancy: HW(2S) + HW(δS)

By duplicating two components and adding a comparison structure, Dual modular redundancy (DMR), or duplex, are common systems to detect errors (von Neumann 1956). DMR uses two fully redundant units working in parallel and has been widely used in low level circuit implementations where a signal is duplicated as an input to two redundant and independent logic gates and it is transparently checked for errors.

Figure 3.8 depicts a DMR structure with a checker component. The checker logic compares the output of block 1 and block 2.

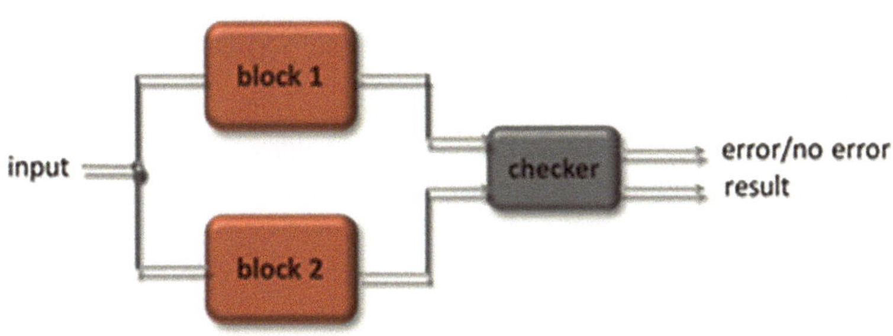

Fig. 3.8 Dual modular redundant (DMR) structure

In the case of normal execution with no error, both blocks would produce the same output and a result would be delivered. On the other hand, in case of a mismatch between the two outputs of the blocks, the output of the checker would produce an error signal and no result will be given.

Therefore, in its simplest version, as the checker logic is unable to identify the incorrect unit, DMR through output comparison will only provide error detection and will not provide error recovery capabilities on its own.

Additional mechanisms will be needed to provide error recovery so that if one of the units experiences an error, the surviving/correct unit can continue execution. Upon successful repair/recovery DMR is fully restored. The only known solution for detecting faulty block in dual scheme was proposed in (Schagaev 1986a, b).

Redundant Execution

A widespread and simple implementation of coarse-grained DMR is lock-stepping, or lock-step execution (Buckle and Highleyman 2003; McEvoy 1981; Sherman 2003). Here, the processor pipeline is duplicated and the clock is shared, comparing each instruction result before committing the results. This type of error detection is considered to perform at the macro-level since it is applied at the microprocessor's scope.

Lock-stepping is widely used in a number of commercial processor designs and can both detect and correct certain errors; e.g. *IBM G5* (Slegel et al. 1999) and Compaq Himalaya (Wood 1999). Redundant threads are executed in multiple processors and every instruction result is compared. No instruction can be committed until its identical pair has also been completed and verified, hence involving considerable overhead.

Stand-by Redundancy

Standby redundancy, standby replacement, or standby sparing, is a well-known fault tolerant design technique used as a failover mechanism (Avizienis 1976). In this case some units are online and operational and one or more backup units serve as standby units.

Figures 3.9 and 3.10 present simple and multiple standby system configurations. When a fault is detected in an online/active unit, a standby unit replaces the affected unit by using the selector (Fig. 3.9) or by using the 3-to-1-switch (Fig. 3.10).

There are three common forms of standby redundancy: so-called *hot*, *warm* and *cold*. The type of application plays a key role in selecting the type of standby spare units. Figure 3.11 graphically describes the typical reconfiguration steps for hot, cold and warm backup spares.

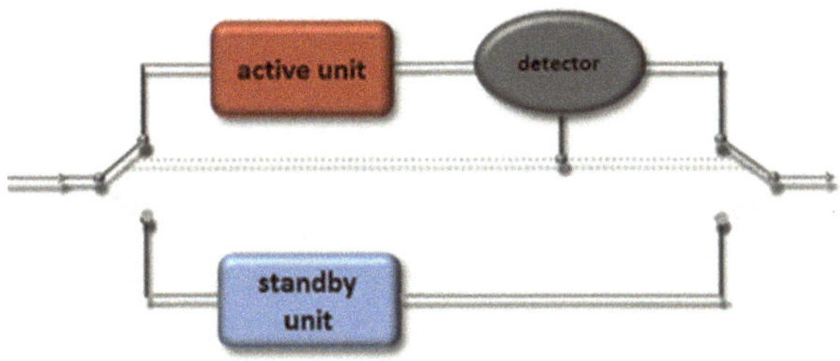

Fig. 3.9 Simple stand-by sparing configuration

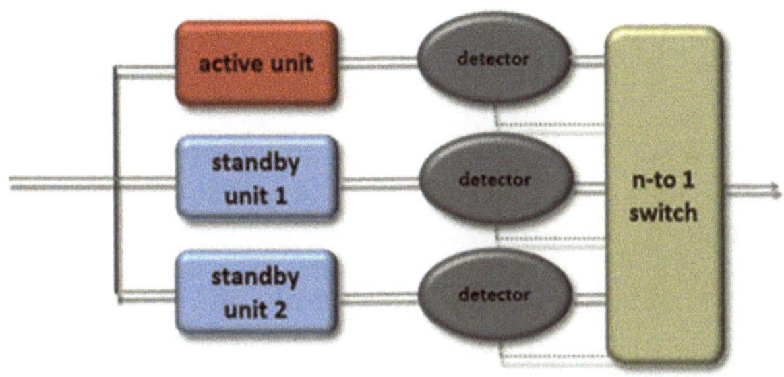

Fig. 3.10 Multiple stand-by spares with n-to-1 switch

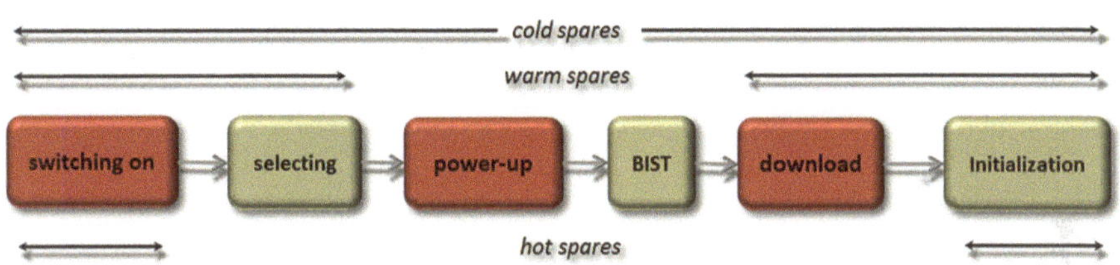

Fig. 3.11 Typical reconfiguration steps for back-up sparing

When a spare unit is to be switched off, the selected spare is powered up and gets ready to become active. The reconfiguration process whereby a standby spare unit becomes operational is composed of:

- Switching on the power and the bus connections
- Powering up of the unit
- Running the *Built-In-Self-Test*[5] (*BIST*): Extensive testing is usually done after powering up to avoid replacing a faulty module with a faulty module before starting normal operation, e.g. memory tests of a spare module
- Loading programs and data
- Initializing the software if needed

[5] Built-In-Self-Tests (BISTs) are one of the common methods of testing circuits. BIST is a DFT technique that takes place on the same substrate as the device under test (DUT) within the system allowing them to perform self-testing (Stroud 2002).

Hot standby spares (*HSP*) operate in synchrony with the operational units and are ready to take over whenever a fault is detected. *HSP* units reduce the mean time to recovery (*MTTR*), and therefore, their use is suitable for applications that require short recovery time, that is, applications where the disruption of processing must be minimised.

Cold standby spares (*CSP*) remain unpowered and thus do not operate or consume any power until they need to replace an active unit. Since the restarting of the units is required, the use of *CSP* is best suited to remote operations where power is hard to come by, e.g. satellites and sensor systems. *CSP* units are also suitable for applications where short lapses in operation are acceptable and state of data is not critical. In addition, *CSP* are likely to have a lower failure rate than operational modules. However, the start-up delay required to switch over to a spare module is high since power up, *BIST* and initialization are needed. In particular, the time necessary for *BIST* depends upon the fault *coverage* and the complexity of the unit/module.

Warm standby spares (*WSP*) consist of a trade-off between the high power consumption of *CSP* and the long recovery time of *HSP*. *WSP* units have time dependent behaviour. Before and after replacing an operational unit, *WSP* present different failure distributions.

The advantage of standby sparing for a system with n identical units is that a certain level of fault tolerance can be provided with $k < n$ spare modules.

3.4.2.2 Pair Spare

The *pair and spare* configurations are a combination of DMR with comparison and extra spare techniques. Figure 3.12 depicts a pair and spare configuration where two units are always online and compared to each other, with any of the n spares being able to replace either of the operational units.

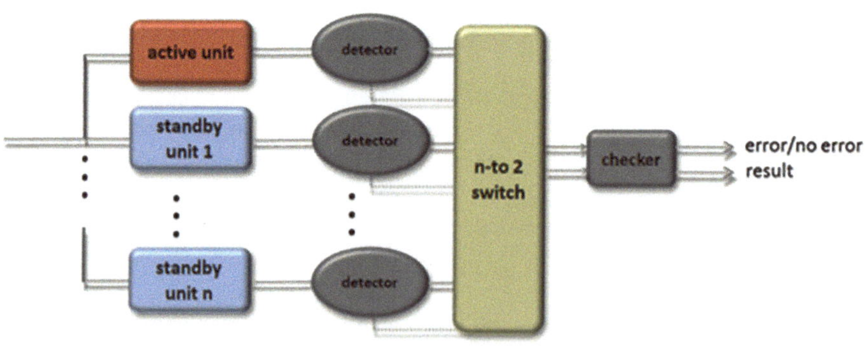

Fig. 3.12 Pair and spare configuration

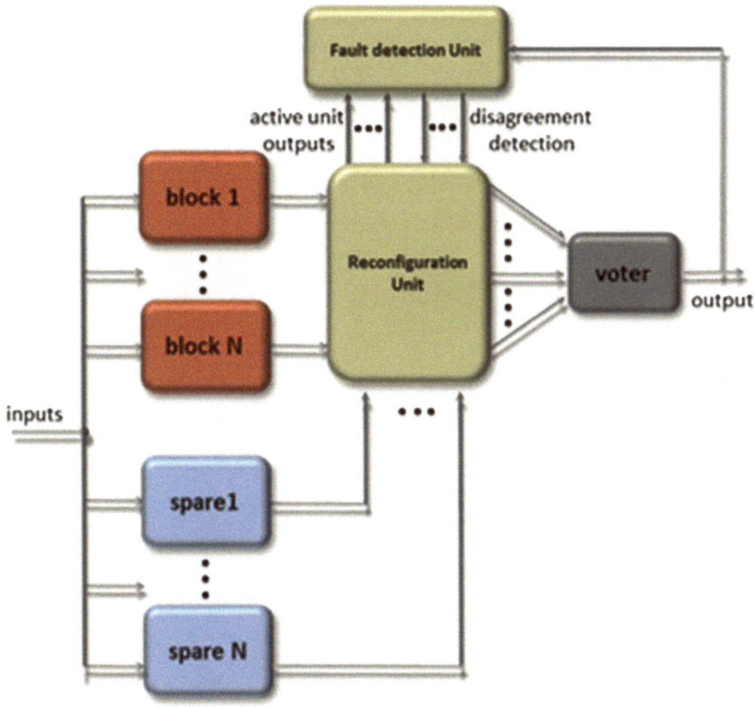

Fig. 3.13 Hybrid approach using TMR with spaces (Johnson 1989)

3.4.3 Hybrid Redundancy

By mixing fault masking, detection location and recovery, the advantages of static and dynamic redundancy can be combined (Johnson 1989). Hybrid approaches use Fault masking to prevent erroneous results from being processed and erroneous data spread across the system. Fault detection, location and recovery are also employed in hybrid techniques to improve fault tolerance by removing errors.

A general approach is to back up the replicated modules with spares, e.g. a TMR configuration with a fault/disagreement detector, a voter and a reconfiguration unit (see Fig. 3.13).

In such a system, the triplicated operational modules are backed up with an additional pool of spares that can replace faulty modules (TMR *with spares*). The system will work as a basic TMR configuration until the disagreement detector determines that a faulty module exists.

One alternative approach towards fault detection is to feed the output of the majority voter back to the faulty detection unit whose job is to compare the output of the voter with the individual outputs of each operational module.

Any disagreement with a specific module's output would indicate that the module should be labelled as faulty and therefore replaced by a spare unit.

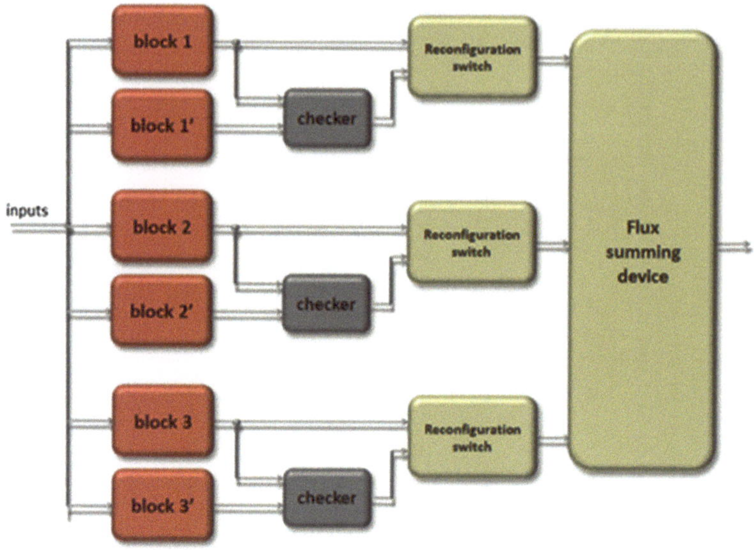

Fig. 3.14 A triple-duplex approach

The reliability of the basic TMR system is retained as long as the pool of spares is not exhausted. Note that voting only occurs among the operational modules in the TMR core, masking faults and making sure that continuous correct service is delivered.

Figure 3.14 presents a variation of NMR with spares is the triple-duplex approach with combination of duplication with comparison and TMR. The use of passive redundancy in the form of TMR allows potential faults to be masked and continuous correct service to be provided for a maximum of two faulty modules. The use of DMR with comparison allows faults to be detected and faulty modules to be removed from the voting process and replaced by spares.

These options are simple but have by far higher overheads than traditional static techniques. Besides, as seen in Section "Reliability of TMR with Voting" the reliability of TMR depends mostly on reliability of applied voters.

Hence, if a fault takes place within a voter, an incorrect majority vote may be given to the output and propagated throughout the system, thus compromising the correctness of the system's service. In order to avoid such unreliability, voters can be designed to be capable of testing themselves online with regards to their own internal faults (Cazeaux et al. 2004; Metra et al. 1997, p. 97).

"Overheads" of redundancy can be effectively used to eliminate malfunctions for classic triple RAM and for relatively new RAM structures with less redundancy (Schagaev and Buhanova 2001).

Fault tolerance of RAM structures can be implemented using already proposed generalised algorithm of fault tolerance (GAFT) (Schagaev 1990) further extended in (Schagaev 2008), (Kaegi and Schagaev 2013).

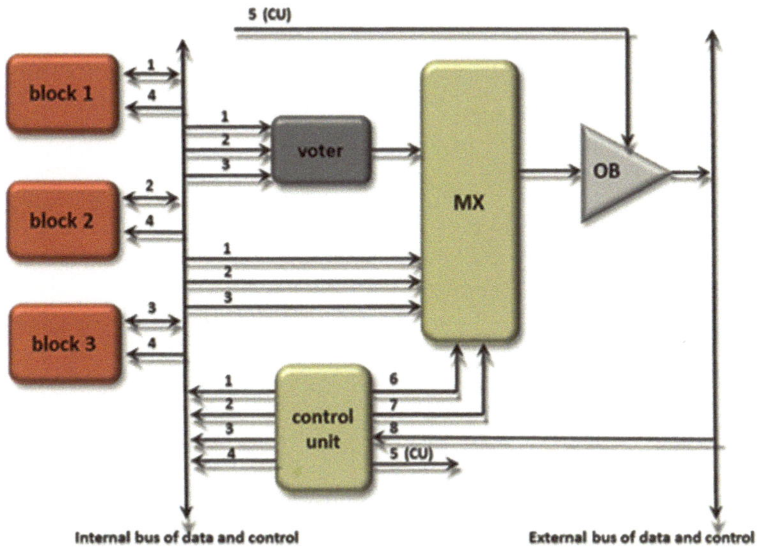

Fig. 3.15 Transient fault tolerant TRAM (Schagaev and Buhanova 2001)

It is well-known that triplicated memory works successfully in terms of timing of operation as fault of hardware manifested like error of information is compensated (masked) in real time of operation and therefore can be invisible for real time safety-critical applications. The only weakness of this scheme is enormous amount of malfunctions that modern electronics suffer due to technological limitations and frequencies used.

Amount of malfunctions in electronics of twenty-first century so far at order of magnitude greater than permanent faults and voting with two malfunctioned devices is no longer possible to ignore. In fact, "library" elements of CAD systems that developed for reliable hardware designs guarantee spreading unreliability! Thus, there is a need to increase TRAM tolerance to malfunctions.

To achieve this, the following solution is proposed. Any working instruction of TRAM (Read, Write) should be accompanied by comparing the context of the same address outputs for all three elements block 1-block 3 (Fig. 3.15).

When the contexts do not match the resulting data formed by standard voting scheme as output signal of TRAM, has to be rewritten to the inputs of all elements, using at the same address. This action takes place when the error is detected, i.e. redundancy is used only when required, causing no delays during regular operations.

Thus, malfunctions affecting the TRAM operation can be tolerated by using existing hardware redundancy (3HW).

This can be considered as an extra form of hybrid redundancy scheme that allows error detection and correction, thus improving the reliability of memory (Fig. 3.15). Any reading and writing operation is followed by a content check of a specific address in all three blocks.

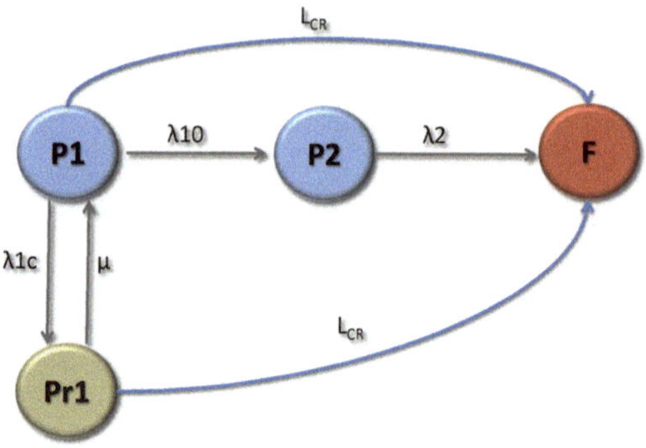

Fig. 3.16 Markov diagram for modernised TRAM

In case of a mismatch among these a majority voting takes place whose result is then rewritten (via control unit) to the inputs of all elements using the same address.

Comments to Figure 3.15: 1, 2, 3 data signals from blocks 1–3 for Read/Write Modes; 4—control signals for blocks 1–3; 5—Signal of control of Output Data Buffer; 6, 7—Signals of control for multiplex to assign mode of operation; 8—Signal of data and control for external buses.

Signals from blocks 1–3 go to voter and after multiplexing go to the output buffer (OB). When malfunction is detected, the correct data will be rewritten to all blocks 1–3—output data will be taken from OB and via lines 8, 4. Control unit performs change of instruction (in fact, rewrite of correct data).

It is worth to consider further modifications of the proposed memory structure, for instance to implement a graceful degradation strategy when permanent faults occur. Working prototype of modernised TRAM was developed using Micron SRAM 32 K by 8 in one DIP package in 1991–1993 and tested in 1994 and still is in very heavy operations without any visible degradation.

Reliability gain for this kind of system can be estimated using Eq. (3.8) which were developed for the Fig. 3.16.

$$\begin{aligned}
&\mathrm{d}P_1(t)/\mathrm{d}t = -(\lambda_{10} + \lambda_{1\mathrm{C}} + \lambda_{\mathrm{CR}})P_1(t) + \mu P_{R1}(t);\\
&\mathrm{d}P_{R1}(t)/\mathrm{d}t = \lambda_{1\mathrm{C}}P_1(t) - (\lambda_{\mathrm{CR}} + \mu)P_{R1}(t);\\
&\mathrm{d}P_2(t)/\mathrm{d}t = \lambda_{10}P_1(t) - \lambda_2 P_2(t);\\
&\mathrm{d}P_F(t)/\mathrm{d}t = \lambda_{\mathrm{CR}}P_1(t) + \lambda_{\mathrm{CR}}P_{R1}(t) + \lambda_2 P_2(t)
\end{aligned} \tag{3.8}$$

λ_{o}—rate of permanent faults, λ_{c}—rate of malfunction, λ_{CR}—rate of fault of control and recovery hardware, $\lambda_{10} = 3\lambda_{\mathrm{o}}$; $\lambda_{1\mathrm{C}} = 3\lambda_{\mathrm{c}}$; $\lambda_2 = 2(\lambda_{\mathrm{o}} + \lambda_{\mathrm{c}}) + \lambda_{CR}$; μ—rate of recovery.

Figure 3.16 presents a diagram of reliability states for TRAM, assuming Markov properties of process of fault appearance and handling. Model has two workable states OK state—P_1, workable state P_2, as well as state of recovery after error P_{r1} and Fatal state F.

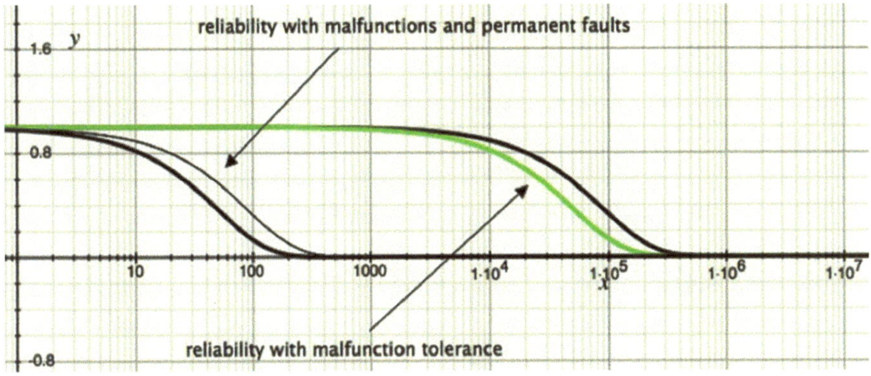

Fig. 3.17 Reliability of modernised TRAM

Without any claim on absolute correctness of these state diagram and equations it is clear that sum of periods when system exist in states P_1, P_2 and P_r is much higher than time of system presence only in the state P_1. This justify efforts required for toleration of malfunction.

Figure 3.17 below compares availability of classic (lower) and proposed triplicate RAM. Time is measured in hours. As it is shown, the acceptable level of availability for the device (0.9) period of operation can be easily increased up to several orders.

Figure 3.17 presents reliability results calculated with the following values: $\lambda_c = 10^{-2}$ (1/h), $\lambda_o = 10^{-4}$ (1/h) and $\lambda_{CR} = 2 \times 10^{-9} -$ (1/h), $\mu = 1$ ms.

3.5 Information Redundancy

Information redundancy involves the addition of new information to existing information i.e. using more bits than needed to ensure fault free functioning. The most common form of information redundancy is *coding* (see Fig. 3.18). Coding theory in hardware and software fault tolerance goes back a very long way and was initially motivated by the need to mitigate errors in information transmission (Shannon 1948).

Coding consists of adding check bits to the data allowing (1) the verification of data correctness and/or (2) the correction of erroneous data. Therefore, with coding an original piece of meaningful information, or *d-bit* data word, is encoded obtaining a *c-bit* code word, where $c > d$ (see Fig. 3.18). Because of these extra bits not all 2^c possible binary combinations are valid code words. Therefore, a code should be selected so that any potential error would transform the codeword, after decoding, into an invalid code word (non-codeword).

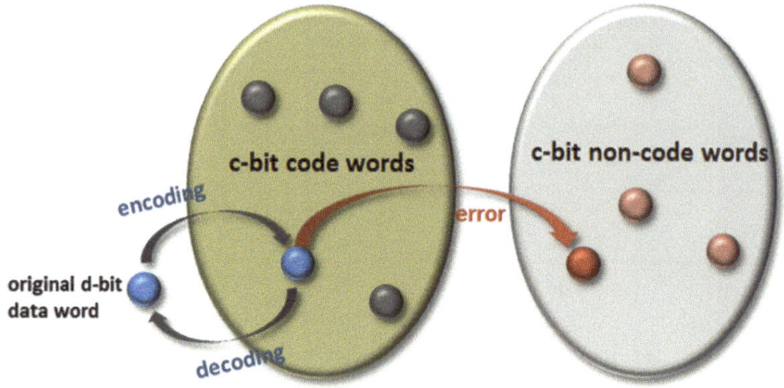

Fig. 3.18 Coding-encoding process of a d-bit word into a c-bit word

An important property of coding is separability. Two main approaches are possible:

- Separable or systematic codes: the code word is formed by adding extra information (check bits) to the original data. A separable code has separable fields for data and check bits. Decoding this type of code is simple and consists of selecting the data bits and disregarding the check bits.
- Non-separable or non-systematic codes: data and check bits are integrated together requiring some extra processing and therefore incurring additional delays and overheads.

Important parameters for codes are:

- The number of erroneous bits that can be detected
- The number of those that can be corrected
- The number of additional bits that are required
- The time needed to encode
- The decoding time

Information redundancy techniques make use of detection-based codes (EDC) or correction based (ECC) codes. Figure 3.19 presents a taxonomy of coding techniques.

Examples of some of these techniques include:

- Error detection and correction codes for cross-checking the contents of main memory, register files and cache
- Cross-checking of run-time control flow using signatures
- Algorithm based checksums for cross-checking of the data values generated

In general, information redundancy involves some space and computational overheads, thus requiring extra circuitry and is thus more commonly implemented in memory structures instead of in processor data-paths.

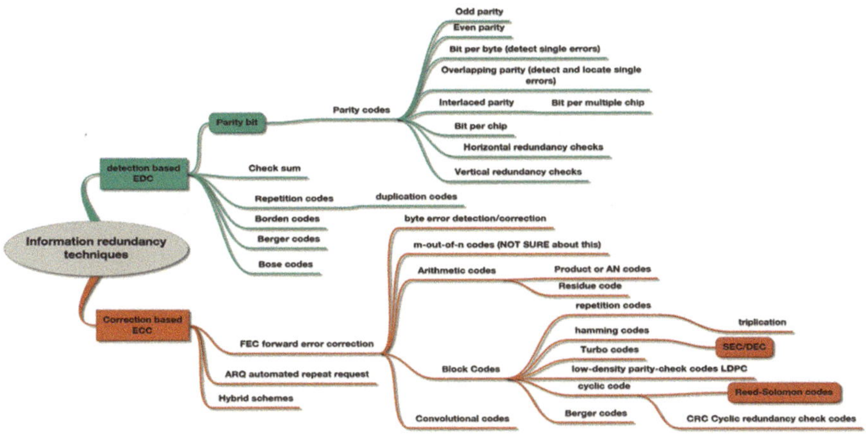

Fig. 3.19 Taxonomy of information redundancy coding techniques

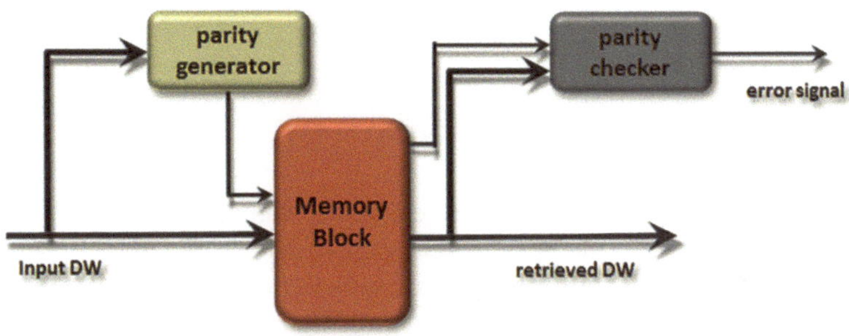

Fig. 3.20 Coding-encoding in memory block with parity checking

3.5.1 *Error Detection Codes: EDC*

Error detection codes have the ability to expose error(s) in a given data word based on the encoding-decoding principles discussed in Sect. 3.5. In general, error-detecting codes (EDC) present less overhead than error correcting codes (ECC) since they do not have correction capabilities.

The simplest EDC are parity codes, which involve the addition of extra bits. Figure 3.20 depicts a basic scheme of memory with parity checking.

Before storing a word in the memory block a parity generator computes the parity bits from the bits of the input data word (DW). A parity bit is an extra bit added to a group of source bits (DWs) in order to ensure that the outcome or coded word has an even (in the case of even parity) or odd (in case of odd parity) number of bits set to 1.

When a memory block is read, the parity checker compares the computed and the stored parity bits, setting the error signal consequently. If both, computed and stored parity bits, match then the error signal would indicate a correct output; otherwise the error signal would indicate that the retrieved DW is incorrect. Note that for n bits of data there are 2^n possible DWs. Adding one parity bit would allow 2^{n+1} possible DWs.

Among these possible *DWs* there are $\frac{2^{n+1}}{2}$ possible DWs with an odd number of 1 s and $\frac{2^{n+1}}{2}$ possible DWs with an even number of 1 s. In the case of odd parity, only the DWs with an odd number of 1 s are valid code words (CWs).

In the presence of single bit flip (error) an odd parity CW would change into an even parity CW and therefore the parity checker will detect the error. Nonetheless, it will not know which bit has been flipped. This simple configuration can be used to detect single or any odd number of errors in the retrieved DW.

However, an even number of flipped bits would make the parity bit of the CW appear to be correct although the data is incorrect. With single parity, double errors and even number of errors would remain undetected. Thus, minimal information redundancy—one bit enables single bit manifestation, but does not provide faulty bit location within word of data or recover of erroneous bit.

3.5.2 Error Correction Codes: ECC

Codes more powerful than parity can be created by adding more check bits to the original data. The size of the data to be protected will determine the number of check bits needed. Using this basic principle, error correction codes have the ability to detect errors and reconstruct the original error-free data. These can generally be realised in three different manners (see Fig. 3.19):

- *Backward Error Correction* (*BEC*) sometimes referred to as Automatic repeat request (ARQ): combines an error detection technique (error detection encoding prior to transmission) with retransmission request of erroneous data. BEC requires simpler decoding infrastructure than FEC but frequent retransmissions would significantly compromise performance in high data rate transmissions.
- *Forward Error Correction* (*FEC*) or Channel Coding: With this approach, errors are both detected and corrected at the receiver's end. Thus, it involves error-correcting encoding prior to transmission without retransmission of the original information. FEC requires more complex decoding infrastructure than BEC but it is suitable for high data rate applications.
- *Hybrid automatic repeat request* (*HARQ*)*:* BEC and FEC are combined, e.g. a scheme where minor errors are corrected without retransmission (FEC) and major errors are corrected via retransmission (BEC).

An overview of popular *FEC* schemes employed in fault tolerant design of embedded systems follows. Figure 3.21 shows a basic ECC memory scheme that applies to any of the following codes including calculation, checking and correcting logic.

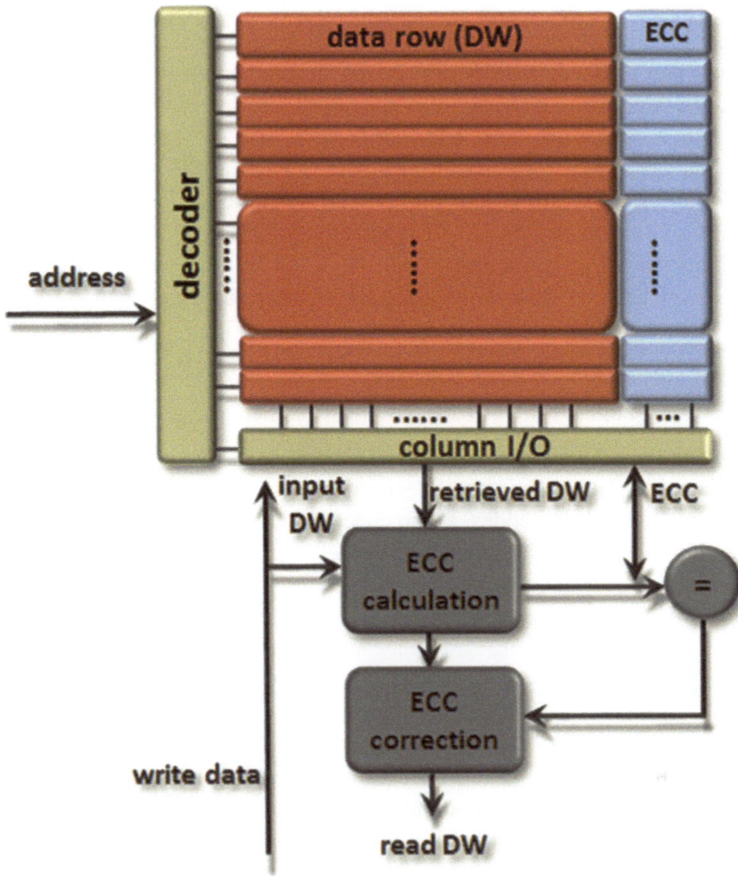

Fig. 3.21 Basic ECC memory scheme including calculation, checking and correcting

When data is written into the data row specified by the address signals, the ECC encoding logic generates the parity checks (as specified by the code) and introduces them into the ECC part of the memory.

When the DW is read from the memory the parity bits would allow missing data to be reconstructed in the case of an error being detected.

3.5.2.1 SEC-DED[6]: Hamming and Hsiao: HW(δI)

The most common ECCs are based on Hamming (Hamming 1950) or Hsiao (Hsiao 1970). These two separable code families introduce the concept of overlapping parity by which every data bit has a part in adjusting the value of

[6] SEC-DED: Single error correction and double error detection.

several parity bits. These codes can correct single bit errors in a given word, can detect double bit errors, are relatively fast decoding and have moderate redundancy. Hamming codes are a family of perfect codes[7] that generalise the original Hamming (7,4)-code (Hamming 1950). A minimum distance d means that it takes d bit changes to move from one valid codeword to the other.

Extended Hamming code sometimes generalised as SEC-DED (single error correction and double error detection), is an example of this type of code. In SEC-DED, an extra parity bit is added achieving a distance of four instead of the three (as in the original Hamming).

The extra parity bit allows the decoder to distinguish between two possible situations:

- When at most one bit flip has occurred
- When two bit flips have taken place

In contrast with *Hamming (7,4)*, SEC-DED provides single-bit-error correction and simultaneous double-bit-error detection.

Compared to Hamming codes, Hsiao codes (Hsiao 1970) provide improvements in speed, reliability and calculation cost as well as checking and correcting logic.

However, in situations that demand higher reliability requirements than those provided by SEC-DED, more complex codes are required.

SEC-DED Limitations and Alternative Techniques

The main limitation of SEC-DED codes is that triple-bit errors may not only remain undetected but it may also be mis-corrected as if they were single-bit-errors (Hsiao 1970). The probability of this type of mis-correction for 32bit data words is around 60 % or more.

Multiple errors are usually taking place in adjacent memory locations, therefore increasing the chances of having multiple bit errors in a given word (Bentoutou and Djaifri 2008; Boatella et al. 2009). These are called *burst errors,*[8] errors that are highly correlated.

If a specific memory cell has an error, it is likely that adjacent cells may also be corrupted by the same event that triggered the error in the first place. Theses are sometimes referred to as *spatial multi-bit errors* (Mukherjee et al. 2004). In contrast, *temporal multi-bit errors* are errors that take place when two different cells of the same word are affected by different events (Mukherjee et al. 2004).

An important risk for SEC-DED schemes is that if a specific memory word is not accessed for a long period of time, the chance of accumulating errors increases (temporal multi-bit errors). In 1983–1987 scheme of reading-rewriting of data from

[7] A Hamming code is perfect in the sense that it can achieve the highest possible rate for codes with a given block length and minimum distance of three (Moon 2005)

[8] Also called cluster of errors.

memory was proposed by (Buhanova G, Schagaev I) excluding an accumulative impact of malfunctions on memory context. Working prototype was developed in 1990–1991 (http://www.it-acs.co.uk/files/itacs_devices.pdf). English version of proposed solution was presented at DSN in 2001 conference and published in (Schagaev and Buhanova 2001).

Later this method was renamed *memory scrubbing* (Mukherjee et al. 2004; Saleh et al. 1990; Weaver et al. 2004), suggesting periodical read of each every memory cell. This may be implemented by having a hardware controller that, during idle periods, reads every memory location searching for errors and correcting any single error found during the process, thus reducing the chance of detected (*DRE*[9] and *DUE*[10]) and undetected errors (e.g. *SDC*[11]).

Scrubbing does, however, require additional SW and/or HW overheads depending on the implementation. In current architectures with high memory bandwidths, HW scrubbing is preferred due to its lower timing overhead. In combination with *SEC*, scrubbing is effective against single-bit and temporal multi-bit errors but not against spatial multi-bit errors.

This problem of spatial multi-bit errors, was solved in 1989 and published in (Bernstein et al. 1992, 1993). It was suggested that logical word of data might be aggregated from keeping bits in separate physical memory elements. Prototype did show exceptional reliability of proposed solution.

Later suggested *memory interleaving* (Haraszti 2000; Reviriego et al. 2010; 2007) was proposed to use in conjunction with ECC ensuring that cells that are physically closely located belong to different logic words. That is, cells that belong to the same logical word are physically apart.

Figure 3.22 illustrates an example of memory interleaving in a four *3-bit* memory word. This type of memory distributes logical data into a non-continuous arrangement. More columns than the number of bits of a single word are added, and the corresponding columns for each word are interleaved.

In this way, *burst errors* are distributed over a number of words each suffering only one single bit error. Any *4-bit-upset* affecting adjacent memory cells would cause four single bit errors in separate words, which can be easily corrected by *SEC-DEC*.

A shortcoming of interleaving is that high interleaving distances (*ID*) involve more complex designs and thus higher area and latency overheads (Baeg et al. 2009;

---

[9] A DRE is a detected recoverable error, a benign type of error since recovery of the normal operation by fault tolerant techniques is possible (Kadayif et al. 2010; Weaver et al. 2004).

[10] A DUE is a detected unrecoverable error. DUE take place when fault tolerant techniques are able to discover and/or report an error, from which recovery is not possible (Kadayif et al. 2010; Weaver et al. 2004).

[11] SDC stands for Silent data corruption. A SDC take place when an error is undetected and causes data corruption (SDC). In this case, the corrupted data could go unnoticed making this type of error benign, or could result in a visible error and/or catastrophic failure such as crashing a computer system (Constantinescu et al. 2008; Kadayif et al. 2010; Weaver et al. 2004).

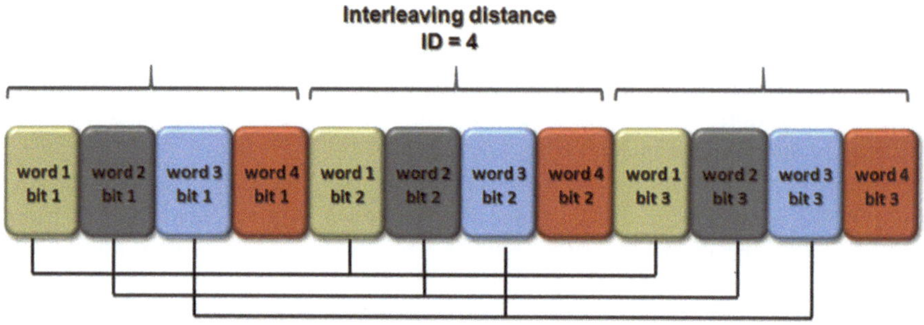

Fig. 3.22 Memory interleaving of four 3-bit words with a 4 interleaving distance (ID)

Reviriego et al. 2010). Ideally the ID should be selected as the maximum expected *MCU* size so that all upsets in a burst error would occur in different logical words.

3.5.2.2 Complex Codes

EDAC implementations based on Hamming codes are the easiest to implement but only provide single error correction (Hentschke et al. 2002). There are alternatives to SEC-DED like Bose–Chaudhuri–Hocquenghem (BCH) (Bose and Ray-Chaudhuri 1960) and Reed–Solomon (RS) codes (Reed and Solomon 1960) based on finite-field arithmetic that can correct multiple faults.

Table 3.4 shows a comparison of the main error correction techniques in memories.

BCH codes are able to correct a given number of bits at any position, whereas *RS* codes group the bits in blocks in order to correct them. *RS* based codes provide a more robust error correction capability but uses a large amount of system resources.[12]

The *RS* decoding process has several stages to get the location of the error and correct it. Implementations of *RS* codes can be found in (Neuberger et al. 2005; 2003).

Although the *RS* algorithm can be simplified (Neuberger et al. 2003) the main disadvantage of these two codes is having complex and iterative algorithms.

As with Hamming based SEC-DED, more complex codes can be implemented based on RS and BCH algorithms. Some examples are SNC-DND[13] (Chen and Hsiao 1984) and DEC-TED[14] codes (Lin and Costello 1983).

Table 3.5 is an overhead comparison of various *EDAC* schemes: Single parity EDC, Hamming SEC-DED, SNC-DND and DEC-TED. Complex errors increase

[12] DEC-TED implementations are expensive from both area-penalty and computational-complexity points of view.

[13] SNC-DND: single nibble error correcting, double nibble error detecting.

[14] DEC-TED: double bit error correcting, triple bit error detecting.

Table 3.4 ECC-TMR comparison

Characteristic	Hamming (SEC-DED)	TMR	RS (DEC-TED)	BCH (DEC-TED)
Area	Small overhead to implement Varies depending on the number of bits (7–32 %)	Extra 200 % plus the voting and correcting logic Number of voters is proportional to the number of units	Varies depending on the number of bits (13–75 %)	Varies depending on the number of bits (13–75 %)
Performance	It can be affected by the coder–decoder functions Proportionally dependent on number of bits to be corrected	High performance. Voter is the only source of delay, hence almost constant delay	Lower performance than *BCH* and much lower compared to Hamming and TMR	Higher performance than RS but much lower than Hamming or TMR
Error Correction	Limited capabilities: it corrects only one single incorrect bit per word.	Corrects up to n errors in an n-bit word as long as the errors are located in a distinct position/unit.	Can handle multiple errors; Efficient for correlated errors (e.g. burst)	Can handle multiple errors; Efficient for uncorrelated errors (e.g. random errors)
Implementation	Binary based Simple to implement	Simple to implement	Symbol based Complex to decode and implement	Binary based Complex but simpler to decode and implement than *RS*

the overhead rapidly as correction capability is increased (Kim et al. 2007). Note also that for any given technique, as the data size increases, the relative overhead of a given scheme decreases (Table 3.5).

Note that the calculation of overheads is just the number of check bits divided by the number of data bits and does not include the extra overheads (e.g. I/O and checkers).

In addition to the area penalty, as the correction capability increases, timing overheads also increase. Results on 64 kb *SRAM* developed in 90 nm processes show that the implementation of a DEC-TED encoder involves a latency penalty of 80–85 % as compared to SEC-DED (Naseer et al. 2006).

Schemes based on information redundancy can also be applied at different levels. For instance, parity codes can be applied to registers, cache and internal memory, whereas SEC-DED can be implemented in external memory, etc. As all these are more complex codes than SEC-DED let alone single parity codes they produce higher overheads as the correction capability increases (Kim et al. 2007).

Thus these codes are not suitable for areas of real-time systems that demand high possessing performance.

Table 3.5 EDC-ECC storage array overheads, based on Slayman (2005)

Data bits	Single Parity		SEC-DED		SNC-DND		DEC-TED	
	Check bits	Overhead (%)	Check bits	Overhead (%)	Check bits	Overhead (%)	Check bits	Overhead (%)
16	1	6	6	32	12	75	11	69
32	1	3	7	22	12	38	13	41
64	1	2	8	13	14	22	15	23
128	1	1	9	7	16	13	17	13

3.6 Time Redundancy

Figure 3.23 presents a list of the most relevant techniques based on time redundancy, which are described in Sects. 3.6.2, 3.6.3, 3.6.4, 3.6.5 and 3.6.6.

3.6.1 Concurrent Error Detection: Basics of Time Redundancy

The main problem with the structural and information redundancy types reviewed is the penalty imposed in the form of extra hardware. At the expense of using additional time, FT techniques based on time redundancy (TR) aim to reduce the amount of hardware required for the implementation.

Time redundancy techniques involve deliberate delay of execution and delivering the results or the re-execution of code using the same piece of hardware and comparing the execution results to determine if a fault has occurred.

This approach was commonly used in the past and is effective in detecting errors resulting from transient faults.

The latest development of use time redundancy for fault tolerance implemented by system software was described in the monograph by Kaegi and Schagaev (2013). Here we mostly concentrate on hardware support and implementation of time redundancy for fault tolerance.

Figure 3.24 shows the basic transient fault detection mechanism based on re-execution. With this technique two or more different computations are performed at different times t_0, $t_0+\Delta t$, and $t_0+n\Delta t$, given $n>1$.

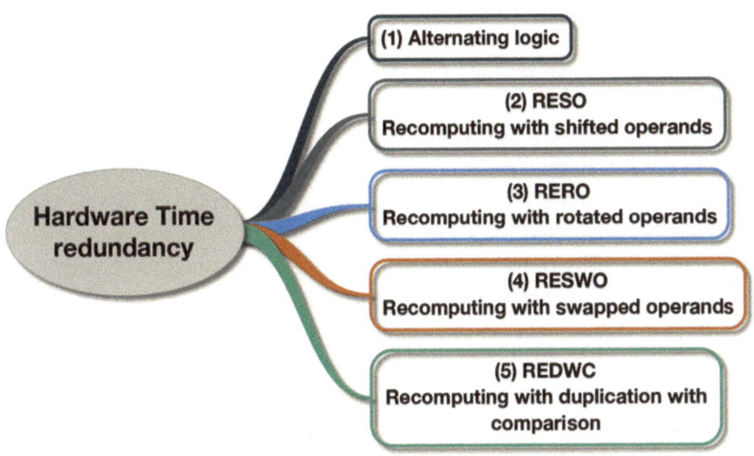

Fig. 3.23 Taxonomy of time redundancy techniques

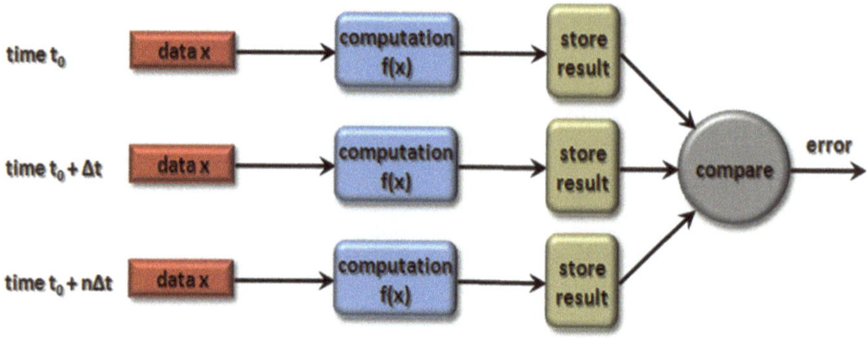

Fig. 3.24 Transient fault detection mechanism based on redundant execution

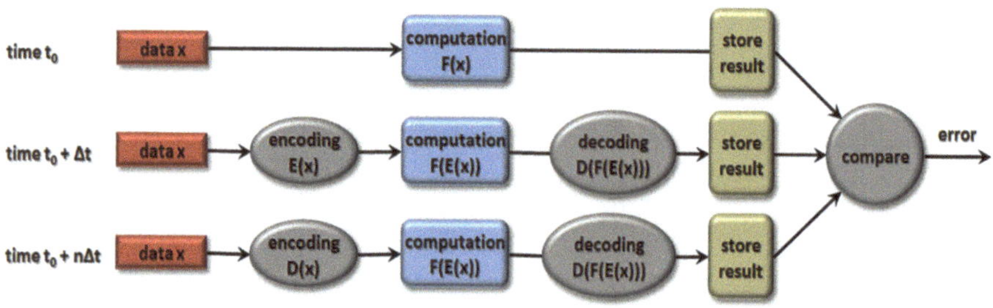

Fig. 3.25 Transient and permanent fault detection mechanism based on redundant execution

The result of a given computation is stored in the corresponding register and then compared to the results obtained from the previous computation(s). If the re-execution is performed twice and a disagreement exists, then transient errors can be detected.

This type of technique was used in the past, but on its own, and did not provide protection against errors due to permanent faults. However, the executions can be performed again to check if the discrepancy remains or not. This is useful in order to distinguish between permanent and transient faults.

If after re-execution the fault disappears, it is assumed to be transient. The hardware resource affected by a transient fault is still usable. On the other hand, if after re-execution the problem persists, the fault is assumed to be permanent and reconfiguration of the specific hardware resource is necessary.

Modern *FT* techniques based on time redundancy can detect permanent faults as shown Fig. 3.25. In this case, during the first computation, the results obtained are simply stored in a register. Then, prior to the next computation(s) a specific type of encoding is performed on the operands. After the relevant computation(s) take (s) place on the encoded operands, the results of all computations are then decoded and compared.

Given that x is the input data, $E(x)$ is the data decoding, $F(x)$ is the functional computation, $F(E(x))$ is the functional computation of the decoding data and $D(E(f(x)))$ is the decoding of the encoded data after computation, and assuming that the functional block is free of permanent faults and assuming that Eq. (3.9):

$$D(E(x)) = x \ \forall x \tag{3.9}$$

Then, the following relation can be stated Eq. (3.10):

$$D(F(E(x))) = F(x) \ \forall x \tag{3.10}$$

If the decoder and the encoding process are carefully selected so that a failure in x would affect $F(x)$ differently than it would affect $F(E(x)$ then if $\Delta t > 0$ the comparison mechanism would produce an error signal.

The main problem with time redundancy techniques is that if the system's data is corrupted by a transient or permanent fault, it will be difficult to repeat a given computation. The critical part of these techniques is assuring that the data is correct and identical before each one of the redundant computations takes place.

The leading concurrent error detection (*CED*) techniques based on time redundancy are *alternating logic*, *recomputing with shifted operands* (*RESO*), *recomputing with rotated operands* (*RERO*), *recomputing with swapped operands* (*RESWO*) and *recomputing with comparison* (*REDWC*). All these techniques are mentioned in Fig. 3.23. The main difference among them is the type of encoding and decoding used.

3.6.1.1 Self-Duality

Self-duality is a property required for certain circuit's functions in order to implement specific error detection techniques based on time redundancy. A function is said to be self-dual if it satisfies the property:

$$f(x_1, x_2, \ldots, x_3) = f\_(x_1, x_2, \ldots, x_3) \ \forall x \tag{3.11}$$

Where x_1, x_2, x_3, …, x_n is the set of inputs to the circuit, $x_{1\_}$, $x_{2\_}$,…, $x_{n\_}$ the set of complemented inputs, $f()$ the output and $f\_()$ the complemented output.

By letting C be a function that complements each bit of a given vector:

$$C(x_1, x_2, \ldots, x_3) = (x_1, x_2, \ldots, x_3) \ \forall x \tag{3.12}$$

It becomes clear that:

$$\begin{gathered} C^{-1} = C \\ C^{-1}(f(c(x))) = f(x) \end{gathered} \tag{3.13}$$

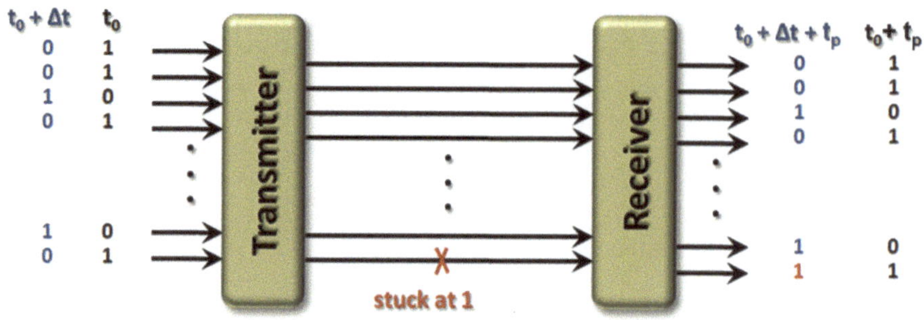

Fig. 3.26 Time redundancy techniques based on alternative logic

Resulting in:

$$C(f(x)) = f(C(x)) \tag{3.14}$$

There are several problems that must be considered when designing a fault tolerant technique using time redundancy. A function C that satisfies the previous property must firstly be determined. Finding a C may not guarantee the desired level of error detection since different circuit implementations based on different C can have different coverage.

Complexity is also an important issue. In the case where the hardware required to implement the coding and decoding functions based on C and C^{-1} is similar to that of implementing $f(x)$, then structural redundancy becomes the more effective choice. In short, the aim of a cost-effective design should be finding a function C that provides a good trade-off between high coverage and low complexity.

3.6.2 Alternating Logic

An example of encoding/decoding function is the complementation operation used in *alternating logic* (Reynolds and Metze 1978) and successfully applied to permanent fault detection of data transmission and digital systems.

As Fig. 3.26 shows the data computed at time t_0 is then complemented and transmitted at time $t_0 + \Delta t$. In the case of a stuck line (either at 0 or at 1) the two computations will generate data that are not complement of each other, and therefore, the error signal will become enabled after comparison.

In this example the last communication line is *stuck at 1*[15], and therefore, complement and data would both become 1, which is not an alternate output, and

[15] A stuck-at fault is a particular fault model used to represent a manufacturing defect within an integrated circuit. Depending on the effect of the fault, a suck-at fault can be stuck either at a logical value of 0 (stuck-at 0) or 1 (stuck-at 1).

therefore, a fault is detected. In order to implement error detection in this coding the circuit function must have the property of *self-duality* otherwise extra input bits would be required. For certain circuits 100 % area (hardware) overheads may be required for certain error detection circuits in addition to time redundancy (Carter and Schneider 1968; Johnson et al. 1988; Woodard and Metze 1978)

The key for fault detection is to determine that at least one input vector exists for which the fault will not result in alternated outputs. Although any single *stuck-at fault* can be detected by this technique, extra redundancy and hardware modifications are required to create self-dual functions from non-self-dual ones. Any non-self-dual function of x variables can be converted into an $x + 1$ variable function that is self-dual and can thus be implemented with an alternating logic circuit.

3.6.3 Recomputing with Shifted Operands (RESO)

Recomputing with shifted operands is a logic level concurrent error detection technique based on time redundancy developed by Patel and Fung (Patel and Fung 1982). RESO can be applied to certain problems in which shifting the inputs forms a complementing function that produces a known relationship in the outputs. It has been originally used for arithmetic and logic units. The error detection capability of RESO depends on the number of shift operations. The generalised version is *RESO-k* and it refers to shifting by *k* bits.

Figure 3.27 shows a schematic of a concurrent error detection mechanism on an ALU using RESO. The operands *a* and *b* undergo a normal ALU operation $f(a, b)$ during the first computation at time t_0 and the result is stored in a register.

During the second computation at time $t_0 + \Delta t$, before entering the ALU the operands are shifted left by *k* bits and the result of the ALU operation is right shifted and finally compared to the ones previously stored in the register. In such operations, left and right shifting can also be denoted as $E(x)$ and $D(x)$ (or $E^{1}(x)$). Therefore, if the equivalent notation for the recomputation is $E^{-1}[f(E(a,b))]$ it should be equal to the first computation $f(a, b)$. If the results are identical the output of the computation will be $f(a, b)$.

However, if there is a discrepancy an error signal will be generated. When an n-bit operand is shifted left by *k*-bit(s), its leftmost *k* bit(s) move out and the right most *k*-bit(s) become zero. This may lead to an incorrect result of $f(a, b)$since k essential bit(s) are removed whenever shifted left.

As with alternating logic, extra redundancy is needed as an $(n + k)$ shifters need to be implemented. Furthermore, a bigger $(n + k)$ bits length ALU is needed, and therefore, the recomputation takes $(n + k)$ bit operations rather than the original n-bit ones.

Furthermore, a totally self-checking equality checker is required for the comparison process and error signalling.

Additionally, parity codes can also be used to detect error in the shifter logic. Note that the fault coverage capability of RESO depends on the number of shifts.

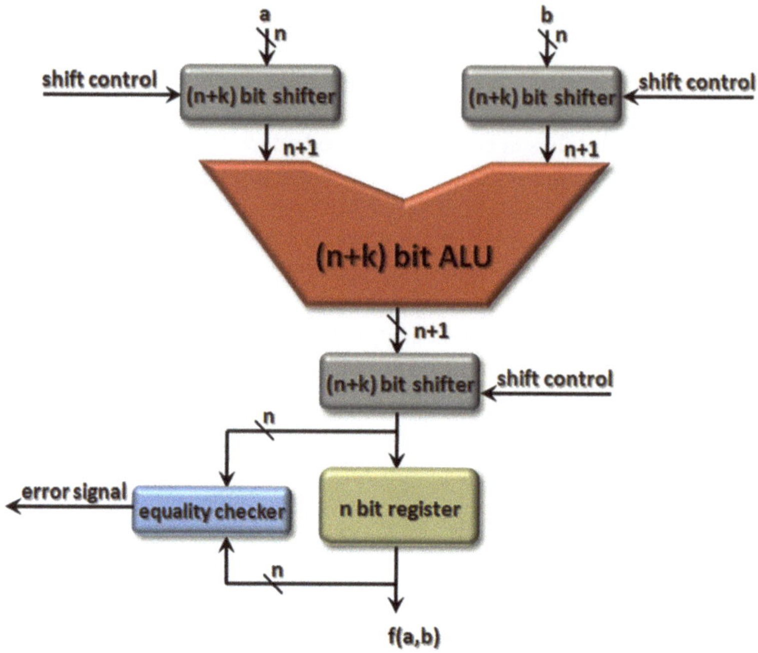

Fig. 3.27 ALU concurrent error detection using recomputing with shifted operands (*RESO-k*)

RESO-1 can detect all single bit-slice errors in an ALU for all bitwise operations, including *AND, OR, NOT, NOR* and *XOR*. As k becomes larger, an increase of space and time complexity is entailed which in turn increase the probability of error.

Consider an ALU with an n-bit shifter and a *RERO-2* implementation and an operand *a* equal to *11010*. After the *11010* is being shifted left by two bits, it will have the two *MSBs* shifted out, thus becoming *01000*. As a consequence, the result of the calculation *f*(*a*, *b*) will probably be incorrect.

If the shifter is replaced by an $(n+k)$-*bit* shifter with $k=2$ in this particular case (*RESO-2*), then the operand *a* after the shifting operation will be equal to *1101000*, thus keeping the *MSBs* and ensuring the correct result *f*(*a*, *b*). Note that during the first computation *k-zero MSBs* are added to each of the operands.

This is one way of detecting errors using RESO. Alternatively, as before, during the first computation at time t_0, the operands *a* and *b* undergo a normal ALU operation *f*(*a*, *b*) but the results are now left-shifted before being stored in the register. In the second computation at time $t_0 + \Delta t$, the operands are also left-shifted by k bits, but in this new way, the results are directly compared with the ones in the register (there is not right-shifting performed on the operands).

The penalty paid for implementing RESO is that every component must be extended to accommodate the shifting. For instance, to implement *RESO-1* on a *32bit* ALU the main system and the shifters are required to be *33* bits, whereas the storage registers and the equality checker must be *34* bits.

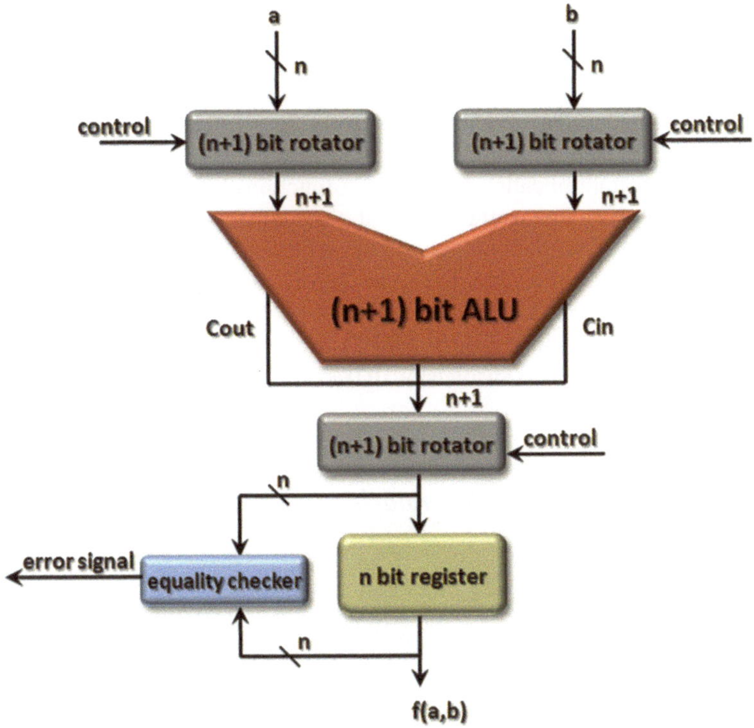

Fig. 3.28 ALU concurrent error detection using recomputing with rotated operands

3.6.4 Recomputing with Rotated Operands (RERO)

Recomputing with rotated operands (Li and Swartzlander 1992) is another technique designed to overcome the limitations of RESO. RERO-k has similar time redundancy characteristics to RESO but with different structural redundancy demands.

Figure 3.28 displays an ALU with RERO-k based concurrent error detection. RESO-k requires an $(n+k)$-bit rotator and an $(n+k)$-bit ALU whilst RERO-k only requires an $(n+k)$-bit rotator and an $(n+k)$-bit ALU.

During the first computation, the $(n+k)$ bit rotators do not rotate the operands, thus the input and output of the rotators is identical. Both operands undergo a regular ALU operation whose result is stored in a register. During the second computations, the first two rotators perform a k-bit(s) right-rotation of the input operands before they enter the ALU. Next, the result is rotated left and compared to the previous result from the first computation. If the results are identical the output of the computation will be $f(a, b)$.

However, if there is a discrepancy an error signal will be generated. With regards to error coverage, a RERO-k implementation with an n-bit ALU can detect:

- $(k\ mode\ n)$ consecutive errors for bit-wise logical operations.
- $(k-1)$ consecutive errors in a ripple carry adder for arithmetic operations.

3.6.5 Recomputing with Swapped Operands (RESWO)

Recomputing with swapped operands (Hana and Johnson 1986) is an extension of the RESO technique that tries to detect errors by alternating the position of the operands. RESWO implementation is very intuitive but with limited applications such as addition, multiplication and Boolean functions but not division or subtraction operations. The first computation at time t_0 is performed on unmodified operands. During recomputation, at time $t_0 + \Delta t$, the operands are first split into two halves, upper and lower, then swapped before calculation and finally swapped back after it. The logic for implementing RESWO has been shown to be less complex and less expensive than in RESO, in particular when the complexity of individual modules is high (Shedletsky 1978).

3.6.6 Recomputing with Comparison (REDWC)

Recomputing with comparison (Johnson et al. 1988) uses a combination of both hardware and time redundancy. The operands a and b of an n-bit operation are split into two halves and computed by two virtually divided devices (n/2-bit size) twice. In a first time slot, the least significant n/2-bits (lower halves) of the operands and their duplicates are carried out and then their results compared. Upon completion, in a second time slot, the same operation is repeated for most significant n/2-bits (the upper halves) of the operands. As long as the separate halves do not become faulty in the same way and at the same time, REDWC can detect all single faults.

3.7 Comparison of Main Redundancy Schemes

In general, the addition of correction capabilities to the error detection mechanisms involves extra area (hardware redundancy) and/or time overheads. Table 3.6 compares the capabilities, timing and area overheads of structural- and time-based *FT* mechanisms.

TR techniques involve low area overheads at the cost of extra timing. Note that some of these techniques, such as RESO, can provide correction capabilities by performing more than two computations. In contrast, at the cost of area penalties, *SR* techniques can provide detection and correction with very little timing overheads.

Error detection codes, such as parity coding involve low complexity and low overheads but have limited detection abilities and are not able to detect multi-bit errors. Although information redundancy schemes can be feasible to correct single and double errors in high-capacity memories (Paul et al. 2011), for $n > 2$, n-bit correction circuitry demands considerable area, energy and timing overheads,

Table 3.6 Comparison of structural-time based FT mechanisms

Scheme	Structural redundancy	Time redundancy	Detection (D) and correction (C)
Time redundancy based			
Alternating logic	≈0–100 %	>100 %	D
RESO	≈0–93 %%	>100 %	D
RESO	≈0–93 %	>200 %	DC
RERO	≈0–93 %	>100 %	D
RESWO	≈0–77 %	0–100 %	D
REDWC	≈0–90 %	0–100 %	D
Structural redundancy based			
DWC	>100 %	≈0–17 %	D
TMR	>202 %	≈0–17 %	DC
TMR with triplicated voter	>208 %	≈0–17 %	DC
Information redundancy based			
Single Parity	1–6 %	≈0–10 %	D
SEC-DED	7–32 %	10–129 %	DC
SNC-DND	13–75 %	–	DC
DEC-TED	13–69 %	22–200 %	DC

especially in low capacity memories. For instance, in an *8-bit* ECC scheme integrated to a 64 kb SRAM the area overhead can be more than 80 % (Kim et al. 2007).

Application of Hamming SEC-DED codes to 16 M-bit *DRAM* chips has a 10 % access time penalty on a to *16 M*-bit *DRAM* (Arimoto et al. 1990; Furutani et al. 1989).

For an experimental 1 *M*-bit *DRAM* cache, applying a SEC-DED code imposes up to 15 % access time overhead (Asakura et al. 1990).

The area penalty is even greater in register files (RFs); experimental results for *SEC* applied to a 64-bit 32-word RF using 90 nm standard cell *ASIC* technology (Naseer et al. 2006) incurs a 22 % area penalty and a 129 % increase in read access time. TMR applied to the same type of registers incurs a 204 % area penalty but increases the read access time by only 17 %. Therefore, for sensitive *ASIC* applications that demand low-latency, TMR is more suitable.

3.8 Conclusion

- The use of fault avoidance techniques does not guarantee complete removal of faults, having many drawbacks in terms of cost, speed of operation and chip area. System testing and verification techniques can never be sufficient to remove all potential faults and their causes.

- In turn, fault tolerant techniques are useful as long as they are applicable along the whole operation cycle of our systems.
- Structural redundancy techniques, such as DWC for single error detection and TMR for single error correction, are very popular. However, both techniques entail high area and power overheads and may not be suitable in embedded applications where power consumption is an important issue.
- CED techniques based on EDC, such as parity coding, involve lower area overheads than structural redundancy techniques but have limited detection abilities and cannot correct errors or efficiently detect multi-bit errors. ECC and physical interleaving incur large area overheads for multi-bit errors. Identifying the time interval for scrubbing can be tricky.
- In terms of time redundancy, the aim of a cost-effective design should be finding a function C that provides a good trade-off between high coverage and low complexity. If the hardware required to implement the coding and decoding functions is similar to that of implementing $f(x)$, then structural redundancy techniques are more effective. Time redundancy cannot be used in every application due to the additional time required. For instance, certain long-life critical systems used in space applications can tolerate additional time much easier than additional space or power requirements, whereas real-time safety critical systems used in avionics cannot afford any additional performance penalty.
- Apart from time, extra hardware in the form of shifters, registers, comparators and extra bits are needed in ALUs. Moreover, fault coverage is not provided for shifters, rotators and comparators weakening our design, unless they are implemented with self-checking capabilities. However, if time is available TR techniques do offer an opportunity to minimise the additional hardware required.
- When it comes to implementing FT, the selection of particular types of redundancy greatly depends upon the application. Therefore, to select a specific set of redundancy techniques for implementation we should examine (a) the different requirements of the specific application and (b) the techniques that are more suitable for such requirements. Likewise, not only the type of redundancy technique is important, but where and at which level it is applied; for instance, applying TMR at the gate, register or circuit level would have a different fault coverage, time, structural and power consumption trade-off.

Chapter 4
Impact of Radiation on Electronics

4.1 Introduction

To develop efficient fault tolerant systems, designers need to be aware of the impact of permanent and transient faults. Hardware faults are a major concern in silicon based electronic components such as SRAM, DRAM, microprocessors and FPGA. These devices have a well-documented history of faults mainly caused by high-energy nuclear particles.

In the cases of safety-critical systems, aerospace and health monitoring systems, maximum reliability can be achieved assuming susceptibility of those systems to faults produced by various internal (e.g. interconnect coupling noise) and external reasons (e.g. cosmic and solar radiation). The traditional reliability analyses of these systems assume failure rates of permanent faults. A typical failure rate for permanent faults due to hard reliability mechanisms such as gate oxide breakdown or metal electro-migration is generally between 1 and 50 FITS.

So far, design and reliability engineers are discounting the effect of transient faults. Moreover, advances in semiconductor technology have been gradually increasing performance. Aggressive scaling of transistor sizes has driven these remarkable improvements in computational performance.

However, the density of modern silicon chips makes them vulnerable to particles of lower energy causing transient faults and, as a consequence, catastrophic failures (Constantinescu 2003; Hazucha and Svensson 2000; Hazucha et al. 2003). Without mitigation mechanisms the error rates due to these transient faults can easily exceed 50,000 FITS per chip.

V. Castano, I. Schagaev, *Resilient Computer System Design*,
DOI 10.1007/978-3-319-15069-7_4

4.2 Radiation and Its Effect on Electronics

The term "radiation" is commonly used to describe a process in which energy travels through a medium, or space, ultimately to be absorbed by another body. Radiation can generally be divided into ionising and non-ionising radiation depending on its ability to ionise matter. Non-ionising radiation does not usually carry enough energy to produce changes to electronic circuitry. Non-ionising radiation can move atoms in a molecule around or cause them to vibrate but does not carry enough energy to ionise atoms or molecules, and as such is not a concern. Non-ionising radiation comes in the form of visible and infrared light, radio waves and microwaves and thermal.

Ionising radiation has enough energy to directly or indirectly remove electrons from atoms or molecules, thus causing the formation of ions. It includes highly energetic protons, alpha particles, heavy ions, galactic cosmic rays and others. Even though neutrons are not ionising particles, their collision with nuclei produces ionising radiation, and therefore, they are also included in this classification.

As manufacturing technologies evolve, the effects of ionising radiation are becoming a primary concern. Semiconductor devices are sensitive to ionising radiation in the space environment, high altitudes and sea levels. There are different radiation damage mechanisms affecting electronics including atomic lattice displacements and ionisation damage. Such mechanisms induce different types of failures such Total Ionising Dose (TID), Single Event Effects (SEEs) and Displacement Damage Dose (DDD).

Resulting particles from distinct radiation sources affect diverse electronic technologies in a variety of ways. Due to the reduction in size of the transistors and the reduction in critical charge of logic circuits, the natural resilience of previous technologies to information corruption is decreasing (Baumann 2002, 2005a, b; Seifert et al. 2002; Shivakumar et al. 2002). Collision of energetic particles with sensitive regions of the semiconductor can alter stored information, potentially leading to logic errors.

Transient faults (Breuer 1973), the predominant faults in modern technologies, can be caused by environmental conditions like temperature, pressure, humidity, voltage, power supply, vibrations, fluctuations and electromagnetic interferences due to crosstalk between long parallel lines in a die. However, ionising particles are the major source of this type of fault.

Transient errors in electronic devices due to ionising radiation in the space environment are well known (Adams and Gelman 1984; Adams et al. 1982; Binder et al. 1975; Blake and Mandel 1986; Waskiewicz et al. 1986) as is the impact of such radiation on application-specific electronics such as commercial (Dyer et al. 1990; Johansson et al. 1998; Olsen et al. 1993) and military (Taber and Normand 1993) avionics, nuclear exposed environments (Mahout et al. 2000; Marshall 1963), medical instrumentation (Bradley and Normand 1998), and other sea level domains (Hauge et al. 1996; Ziegler 1996).

Bird view, or taxonomy on short-term impact of radiation in silicon-based electronics mostly covered in this chapter is presented on Fig. 4.1.

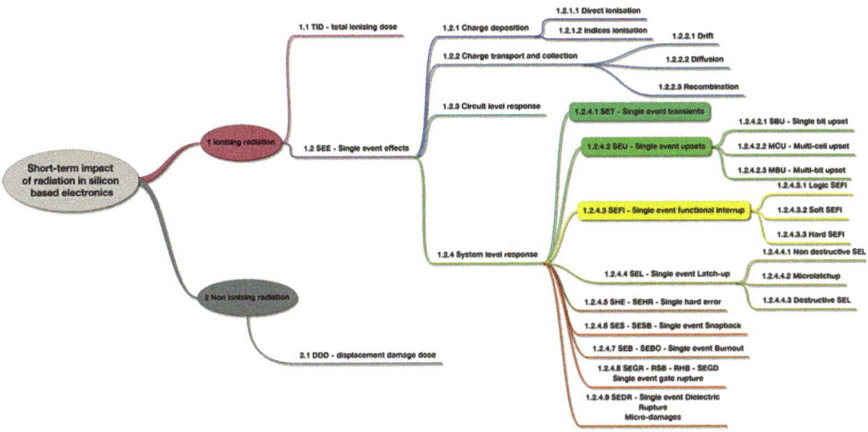

Fig. 4.1 Taxonomy of radiation effects in silicon based electronics

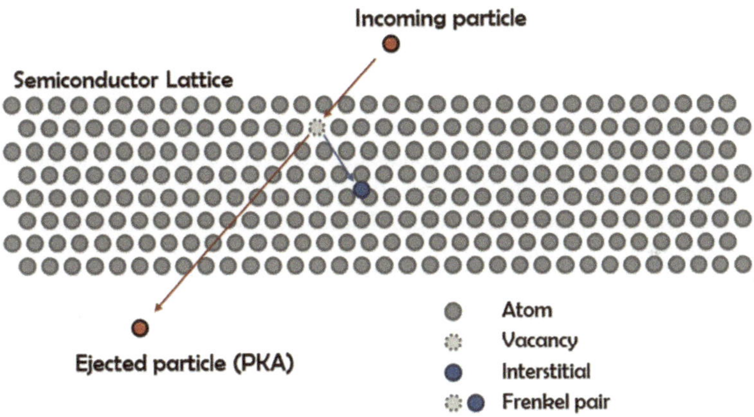

Fig. 4.2 Atomic lattice displacement

4.3 Damage Mechanisms

The two fundamental damage mechanisms (Claeys and Simoen 2002) to electronic elements due to radiation are atomic lattice displacements and ionisation damages.

Atomic lattice displacement occurs when an energetic particle undergoes a nuclear collision with one or more atoms of the electronic device, changing its original position (see Fig. 4.2 above) and thus the analogue properties of the semiconductor junctions, potentially worsening in the long term the properties of the material and creating lasting damage.

In silicon, an impacted atom can become displaced if it is part of the crystalline structure and the incident particle is capable of inducing a minimum energy (displacement threshold energy) of around 20 eV (Miller et al. 1994).

The displaced atom is referred to as "primary knock-on atom" (PKA) and its new non-lattice position is called "interstitial", while its absence from its original lattice position is named "vacancy". Normally, the simplest configuration is a vacancy and an adjacent interstitial generated as a result of a low energy particle hitting the material, a combination referred to as "Frenkel pair".

However, in most cases the displaced atom has enough energy to knock out a neighbouring atom creating a more complex configuration called "cluster", altering the properties of the bulk semiconductor material. In silicon, vacancies and clusters are of unstable nature and tend to be filled by near atoms leading to more stable defects.

In general, this migration leads to the most typical process, called "defect reordering" or "forward annealing" reducing the amount of damage and its effectiveness. Yet, in some cases, depending on the time, temperature and nature of the device, "reverse annealing" can take place, resulting in more efficient defects.

Ionisation damage is primarily induced by charged particles usually leading to transient effects causing temporary variation of the functionality of the system. Since no permanent damage is induced in the electronic circuit, this type of error is called soft error. Ionisation damage may also lead to small degradation and permanent errors, also called hard errors.

A key factor in the damage process is the critical charge, or Qcrit, which is the smallest amount of charge that can cause a change of value in a cell. The effects provoked by the above damage mechanisms can vary depending on the type or combination of types of radiation, radiation flux, total dose, critical charge of the device and manufacturing technology. These factors make modelling of faults difficult and time consuming.

4.4 Radiation Macro-effects

Three major macro effect categories may be used to classify the resultant effects: Total Ionising Dose (TID), Displacement Damage Dose (DDD) and Single Event Effects (SEE). As far as the type of degradation that these macro effects have, TID and DDD are considered as long term cumulative and SEE as short term. Table 4.1 summarises the characteristics of these radiation macro effects.

Total Ionising Dose is a measure of the cumulative effects of the prolonged exposure to ionising radiation. In the context of silicon devices, it is also called surface damage. MOS and bipolar electronic technologies are affected by TID and once the material is damaged, it will not return to its original state (Felix et al. 2007).

In today's devices, the formerly used bipolar transistors have been almost completely replaced by the MOSFETs (Metal Oxide Silicon Field Effect Transistors). The schematic of a typical MOS transistor is shown in Fig. 4.3. Its basic architecture is based on an N-(P-) doped silicon substrate and two highly P-(N-) doped contacts, the source and the drain. The gate oxide covers the channels between the source and the drain.

Table 4.1 Characteristics of radiation macro-effects

Radiation effect	Type of degradation	Source	Damage mechanism	Microeffects	Counter measures mitigation techniques	Sensitive technologies	Temperature dependency
Total ionising dose (TID)	Long-term cumulative	Trapped protons, trapped electrons and solar event protons	Ionising damage	Small energy transfers deposited uniformly and delivered over a long time	• Partial mitigation: Additional shielding is only effective in particular technologies and environments Robust electronic design. High drive currents. High noise immunity, large gain margins, etc. Cold redundancy using spares. Not suitable for all technologies.	Power MOS, CMOS, NMOS, PMOS, SOI, SOS, Bipolar, BiCMOS	Yes
Displacement damage dose (DDD)—Bulk damage	Long-term cumulative	Trapped and solar protons and neutrons	Atomic lattice displacement damage	Accumulation of small energy transfers to atomic nuclei (Coulomb, nuclear interactions)	• Shielding is not only ineffective, but it is also the root of the problem	Bipolar, BiCMOS	No

(continued)

Table 4.1 (continued)

Radiation effect	Type of degradation	Source	Damage mechanism	Microeffects	Counter measures mitigation techniques	Sensitive technologies	Temperature dependency
Single event effects (SEE)	Short-term	GCRs, particles from solar events, trapped protons, and secondary neutrons	Ionising damage	Sudden large energy transfers at the wrong place and time	• Additional shielding is NOT effective • Ensure systems are not sensitive to transient effects • Fault tolerant design techniques • Error detection and correction for critical circuits • System autonomous re-boot	Power MOS, CMOS, NMOS, PMOS, Bipolar, SOI, SOS, BiCMOS	Yes

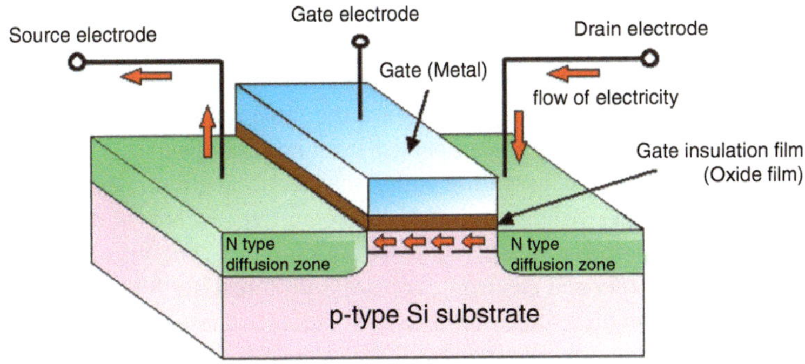

Fig. 4.3 Schematic of MOS transistors

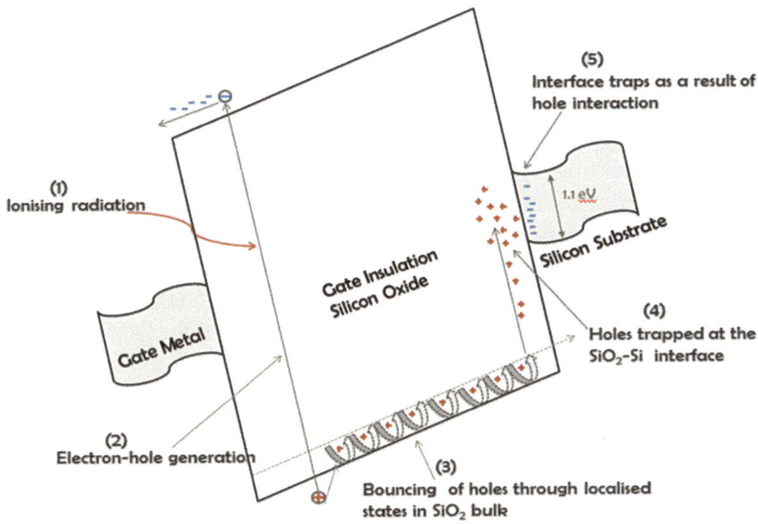

Fig. 4.4 Schematic of the motion of electron holes in a silicon oxide

This thin silicon dioxide (SiO_2) insulating layer is situated under the gate electrode and can attract charge carriers into the channel region. If no voltage is applied at the gate electrode, no current can flow between drain and source. By regularly applying low voltages at the gate, the current between drain and source is regularly switched on and off.

When a highly energetic particle strikes the semiconductor material, as shown in Fig. 4.4, electron hole pairs are generated but disappear quickly due to the low resistance of the gate and the substrate.

However, in the oxide, and due to their different mobility, electrons rapidly move either to the gate or to the channel whereas the holes slowly bounce from site to site until they become trapped[1] by defects near the silicon oxide interface. Some of these holes may be trapped for a long time resulting in a positive charge in the oxide that can affect the characteristics of the transistor and generate shifts in its operating threshold. These voltage shifts are the most common form of radiation damage in MOS technology and can persist from hours to years.

TID effects can lead to degradation within the electrical circuit (threshold shifts), decreased functionality, switching speed, device current, increased device leakage (higher power consumption) and even functional failures. The primary sources of TID are trapped protons and electrons, and solar protons (Barth et al. 2004).

Modern submicron electronics offer relative relief to these effects in the way of natural radiation hardening (Pouponnot 2005; Velazco et al. 2007). Current gate oxides are around 100 times thinner than the approximately 100 nm oxide layers employed in the early 1990s. Modern gate oxides are around 1 nm thick, which allow electrons to tunnel through the potential barrier at the silicon oxide interface, neutralising the trapped holes. Since there is not enough trapped charge, transistor threshold shifts cannot be generated.

Circuit level radiation hardening techniques, i.e. changes in the geometry of transistors, have been used to mitigate TID effects but such techniques are expensive. Furthermore, TID effects might be partially reduced with the use of shielding material that absorbs most electrons and low energy protons. However, the amount of shielding is inversely proportional to its effectiveness in stopping the protons with higher energy (Dyer et al. 1996). TID is considered a severe problem (Claeys and Simoen 2002) during the lifetime of satellites.

Displacement Damage Dose (DDD) or "Bulk" damage (Barth et al. 2004; Yu et al. 2005), occurs when high energy particles dislodge or displace atoms from the semiconductor lattice due to its long time exposure to non ionising energy loss (NIEL). DDD results in a similar long-term cumulative degradation to that caused by TID.

The damage mechanism is the result of collisions with atoms, which become displaced from the lattice creating interstitials and vacancies. Consequently, DDD is an effect of concern for all semiconductor bulk based devices such as bipolar devices (BJT circuits and diodes), BiCMOS, electro optic sensors (CCDs, photodiodes, phototransistors), silicon detectors and solar cells, whereas CMOS is almost insensitive to it.

DDD accumulation primarily occurs when the semiconductor material is exposed to neutrons, trapped protons and solar protons over time. Likewise, secondary radiation produced in shielding materials can cause DDD effects. The overall effect of DDD in semiconductors is alteration in the minority carrier lifetimes, which results in lower currents between the collector and the emitter and therefore reduced transistor gain. An extended review of literature related to this type of damage can be found on (Srour et al. 2003).

[1] In MOS structures oxide traps are defects in the SiO_2 layer, interface traps are defects at the Si/SiO_2 interface and border traps are defects near the interface (Fleetwood et al. 2008).

4.5 Single Event Effects (SEE)

The term Single Event (SE) is used to lay emphasis on the fact that the effect is caused by an individual particle interacting with the material. In current semiconductor technologies single event effects represent a much larger problem than the combination of all long-term cumulative effects.

SEEs are induced by the strike of a single energetic particle (ion, proton, electron, neutron, etc.) in sensitive regions of the material.

The particle travels through the semiconductor material leaving an ionised track behind depositing sufficient energy to cause an effect on a localised area of the electronic device. Both TID and SEE take place as a result of ionising radiation; however, whilst the former is a long term effect that changes the electrical properties of the device, SEEs are the result of an instantaneous perturbation.

Neutron and alpha (α) particles are the most common sources of SEEs in terrestrial environments whilst cosmic rays and heavy ions are most responsible for space applications. SEEs affect many different types of electronic devices and technologies resulting in data corruption, high current conditions and transient disturbances. If not handled well, unwanted functional interruptions and catastrophic failures could take place.

4.5.1 Physical Mechanisms Responsible for SEEs

In the "technological shrink model" of sensitivity to upsets (Baumann 2002; Seifert et al. 2002; Shivakumar et al. 2002) the detailed physical mechanisms responsible for SEE are identified in four consecutive steps (Dodd and Massengill 2003; Wirth et al. 2008) taking place before an SEE occurrence:

- Prior charge deposition by the incident particle striking the semiconductor,
- Transport of the released charge into the device,
- Charge collection by the different sensitive regions,
- Circuit response.

4.5.1.1 Charge Deposition

Ionising radiation can release charge in the semiconductor in different ways. SEEs can occur through the impact of the incident particles themselves (e.g. direct ionisation from galactic cosmic rays (GCRs) or solar particles). SEEs can also occur as a result of secondary particles generated via inelastic or elastic nuclear reactions (Howe et al. 2005; Reed et al. 2006; Warren et al. 2005) and Coulombic (Rutherford or inelastic Coulomb) scattering (Wrobel et al. 2006) between the incident particles and the stationary targets in the struck material (indirect ionisation).

An incident particle can experience a number of interactions before its kinetic energy is expended. In every interaction the path of the particle can be altered and can lose some of the kinetic energy. To measure the energy transferred to the material the terms Linear Stopping Power and Linear Energy Transfer (LET) can be used. Equation (4.1) describes the rate at which a particle loses energy while moving through an absorber. The incremental energy (*dE*) may be expressed in units of MeV, while the path length (*dx*) may be expressed in units of cm.

$$S(E) = -\frac{dE}{dx} \tag{4.1}$$

From these interactions, two types of stopping power can be distinguished (ECSS 2007; Podgorsak 2009):

- Nuclear stopping power (also called radiation stopping power) resulting from energy loss per unit path length due to inelastic Coulomb interactions between the charge particle and the nuclei of the absorber. Only light particles, such as electrons and positrons, experience significant energy loss via nuclear stopping power. For heavier charged particles, such as protons and α particles, this type of loss is insignificant.
- Electronic stopping power (also called ionisation or collision stopping power) resulting from inelastic Coulomb interactions between the charge particle and orbital electrons of the absorber. Electronic stopping power describes the energy lost due to direct ionisation. Unlike nuclear stopping power, heavy and light particles experience this type of interaction that results energy transfer from the incident particle to the orbital electrons via excitation and ionisation (ECSS 2007).

The electronic, nuclear and total stopping energy of different particles are presented Fig. 4.5 (for protons) and Fig. 4.6 (for electrons). Figure 4.5 shows that at all energies the electronic stopping power of protons dominates and that the nuclear stopping power is insignificant. Figure 4.6 shows that the nuclear stopping power of electrons dominates at higher energies.

The total stopping power $S(E)_{tot}$ for a charged particle with E_k energy passing through an absorber of atomic number *Z* is in general the sum of nuclear stopping power and electronic stopping power as shown in Eq. (4.2) (Podgorsak 2009):

$$S(E)_{\text{tot}} = S(E)_{\text{nuclear}} + S(E)_{\text{electronic}} \tag{4.2}$$

Charge deposition is often characterised by mass stopping power, instead of Linear stopping power. Mass stopping power is defined as the Linear Energy Transfer (LET) (not equal to Linear Stopping Power, but approximated) and can be obtained by dividing *S*(*E*) (expressed in MeV/cm) by the density of the material *p* (expressed in mg/cm^3). (Nearly independent of the density of the material, LET) (Eq. 4.3) describes the linear rate of energy transfer to the material as the energetic particle traverses the absorber.

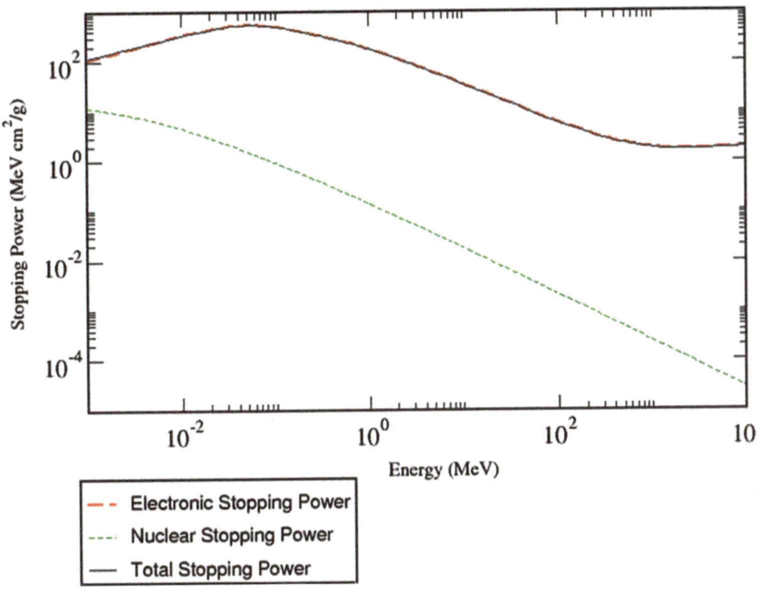

Fig. 4.5 Electronic, nuclear and total stopping power of protons in silicon, computed with PSTAR from NIST laboratory (Berger et al. 2005)

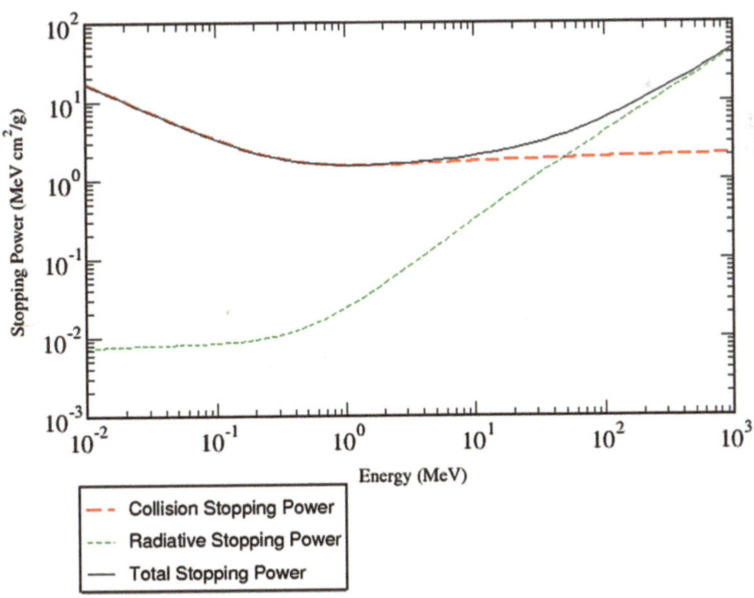

Fig. 4.6 Electronic, nuclear and total stopping power of protons in silicon, computed with ESTAR from NIST laboratory (Berger et al. 2005)

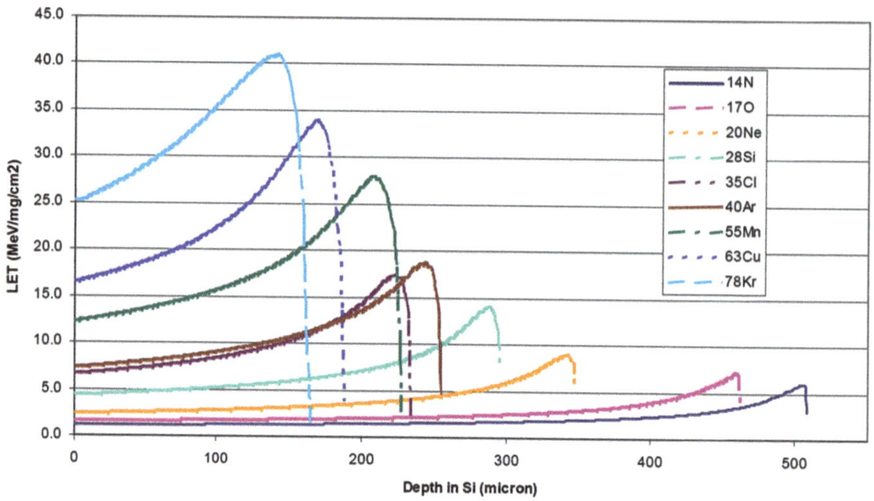

Fig. 4.7 Bragg peaks: LET (MeV/cm^2) of the standard components of a 16 MeV/nucleon cocktail versus depth in silicon (μm) (McMahan et al. 2004)

$$\mathrm{LET} = \frac{1}{p}\frac{dE}{dx} \tag{4.3}$$

The LET of an incident ion and thus, the density of ionisation, typically increases to a maximum immediately before the particle comes to rest. This peak, the Bragg peak, occurs due to the increasing cross section as the particle loses energy.

Figure 4.7 shows a plot of the LET of the standard components of a 16 MeV/nucleon cocktail as a function of depth in silicon.

The LET of a given ion is dependent on its energy and the target material, and therefore is an important parameter to quantify the sensitivity of electronic devices.

Theoretical and experimental values of LET for most ions in different materials have been published (Northcliffe and Schilling 1970). In addition, stopping power for different particles can be calculated using the TRIM code (Ziegler et al. 2010), and the ESRAR, ASTAR and PSTAR programs (Berger et al. 2005).

The LET can be converted into charge per unit length (fC/μm or pC/μm). This is more suitable to situations that take into account the physical dimensions of the device and the charge stored at the critical nodes. For example, in silicon based technologies a particle with a LET of 97 MeV-cm^2/mg corresponds to a charge deposition of approx. 1 pC/μm.

In passing through a semiconductor material, high energetic particles (direct ionisation Fig. 4.8a) can deposit energy in the absorber through a one step process involving Coulomb interactions with the electrostatic field electrons in the target atom (Podgorsak 2009).

The energy introduced allows bound electrons to leave their atoms, releasing free electron hole pairs and converting their energy into charge (Fig. 4.8b).

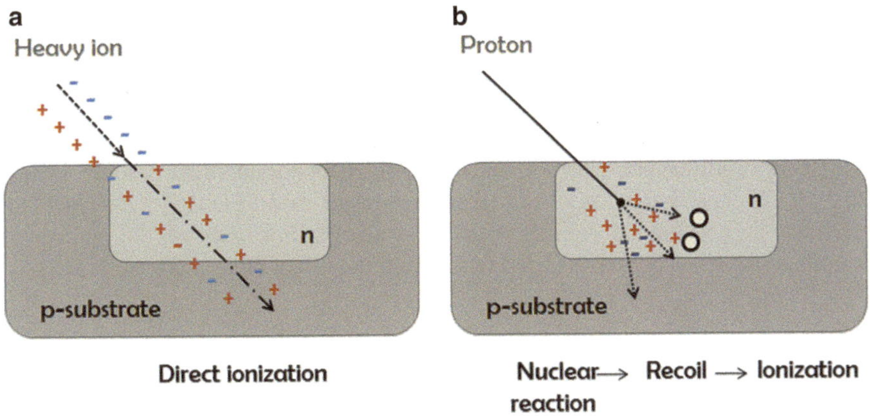

Fig. 4.8 Energetic particle strike and generation of electron hole pairs: (**a**) direct ionisation due to heavy strike; (**b**) indirect ionisation due to proton strike

The particle rests in the semiconductor material once almost all its energy is lost. The energy lost due to direct ionisation can be referred as the electronic stopping power.

The total path length or total distance travelled is referred as particle's range and is highly dependent on the type of particle, its initial energy and the properties of the semiconductor material.

At sea level, direct ionisation is the main charge deposition mechanism for upsets caused by heavy ions and alpha particles, emitted due to the contaminants in packaging materials. Traditionally, since protons and neutrons are lighter, the charge released by them is not enough to produce upsets via direct ionisation.

As suggested in 1997 the technological shrink model would soon be affected by direct ionisation of low energy particles (Duzellier et al. 1997). Recent experimental evidence (Heidel et al. 2008) of 65 nm SOI SRAM sensitivity to direct ionisation from protons supported the latter suggestion with results that the low energy proton for the 65 nm technology is different to those from previous generations.

However, the most significant upset rates due to light particles are caused via indirect ionisation mechanisms. In fact, in today's semiconductor technology, high-energy neutrons derived from cosmic rays are the primary contributor to soft error rates at sea level. In those mechanisms, the highly energetic particles (protons or neutrons) do not directly interact with the material. The three indirect ionisation mechanisms are:

- Inelastic nuclear reactions that take place when the incident particle hits a target nucleus causing fragmentation and ejection of secondary particles;
- Elastic nuclear reactions that take place when the incident particle transfers some of its energy to a target nucleus that recoils (Fig. 4.8) with extra energy transferred from the incident particle;
- Coulombic scattering, similar to elastic nuclear reactions, takes place when the incident particle gets close to a target nucleus that recoils due to Coulomb force with less momentum and smaller angle than with elastic nuclear reactions.

Among these three mechanisms, inelastic nuclear reactions have the higher probability of depositing larger amounts of charge, and hence are the most significant indirect mechanism in the formation of SEE. If an inelastic nuclear reaction takes place, a collision with a target nucleus leads to the emission of reaction products that can, in turn, deposit energy via direct ionisation.

Those resulting particles are much heavier than the incident particle, which involves higher charge deposition that may result in a SEE. Since the incident particles do not directly interact with the semiconductor material, the number of counts or neutrons per cm^2 is used to measure the effect rather than the LET.

4.5.1.2 Charge Transport and Collection

Subsequent to the charge deposition, the released carriers are transported and collected by the semiconductor elementary structures. The transport of the charge is based on three main mechanisms (Dodd 2005):

- Charge collection by drift: The charge can drift in regions with an electric field. Reverse biased semiconductor p-n junctions are usually the most sensitive regions. If the ionised track affects one of those junctions, the high electrical field present in the region can collect the incident charge, which can result in significant transient currents. This is a fast mechanism in the order of 100 ps.
- Charge collection by diffusion: the charge may diffuse in neutral zones (bulk of the device), leading to considerable transient currents. This is a slow mechanism in the order of nanoseconds.
- Recombination: The charge can recombine with free carriers in the lattice.

As Fig. 4.9 illustrates the charge collection can be extended via "field funnelling" (Hsieh et al. 1981, 1983). If a high field region, such as the depletion region of

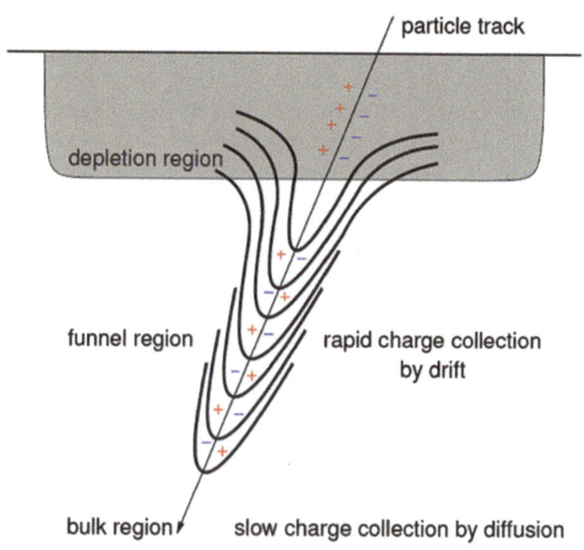

Fig. 4.9 Funnelling effects and charge collection mechanisms (Messenger and Ash 1992)

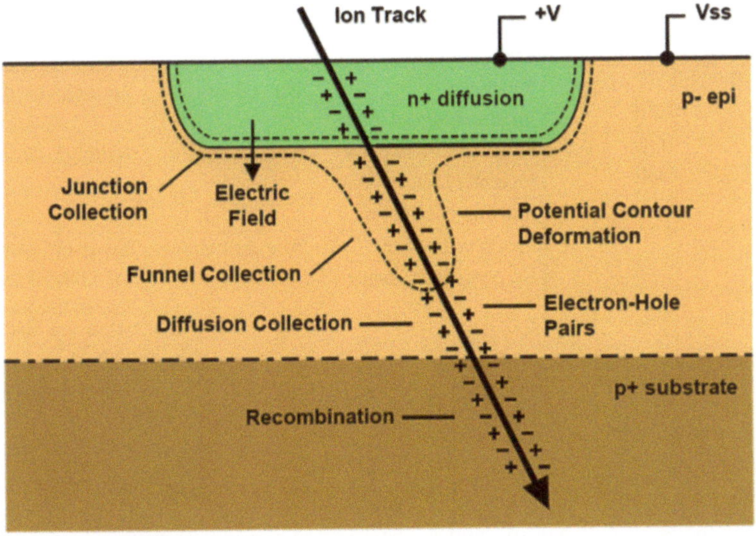

Fig. 4.10 Funnelling effects and charge collection mechanisms after a particle strike on a p-n junction (Mavis 2002)

a p-n junction, is traversed by a column of electron holes, the associated electric field can be disturbed, spreading down along the particle's track deep into the substrate, consequently reducing the net charge in the depletion region.

Three different areas within the track can be distinguished: (1) the initial depletion region, (2) the funnel region and (3) the bulk region. Within the external depletion region, positive potential areas attract the electrons and negative potential areas attract the holes.

Rapid collection by drift will take place in the funnel region whilst the diffusion mechanisms will slowly collect the charge of the residual carriers in the bulk region. The "funnelling effect" is effective in the range of a few nanoseconds.

The generated carrier density in the vicinity of the junction becomes similar to the substrate doping concentration, and the electrical field is then re-established back to its original position as figure below illustrates (Fig. 4.10).

4.5.1.3 Circuit Level Response

The collected charge transported in the device induces parasitic transient currents, which turn could induce disturbances in the external circuits. Depending on (a) the collected charge, (b) the intensity of the resultant current transient, (c) the details of the circuit application and (d) the area affected, the excess of charge can be manifested as one of many types of SEE (or a combination of them).

Semiconductor devices experience SEEs in two major forms: in the form of destructive effects, which result in permanent degradation or even destruction of the

Table 4.2 Type of errors and how to fix them

Type of error		Characteristics	Nature	Fix
Soft	Transient soft	Functionality in place Incorrect logical value	Non-destructive	Reading or writing
	Firm, Static soft	Functionality in place Incorrect logical value Reading does not fix it	Non-destructive	Writing
Pseudo-hard		Functionality lost No permanent damage	Non-destructive	Power-off cycle or reducing the power supply voltage below the holding voltage
Hard		Functionality lost Physical-permanent damage	Destructive	Replacement of HW

device affecting functionality, and in the form of non-destructive effects, causing no permanent damage. Table 4.2 presents different type of errors, their nature, characteristics and a solution to eliminate their effect.

Soft errors are of temporal nature and imply that the physical functionality of the circuit is not affected even though its temporal integrity is. Soft errors have been defined (www.jedec.org/sites/default/files/docs/JESD89-3A.pdf) as "*an erroneous output signal from a latch or memory cell that can be corrected by performing one or more normal functions of the device containing the latch or memory cell*".

Typical examples of this are undesired changes of logic value in sequential logic and undesired analogue pulses that temporarily change the output of combinational logic. Soft errors can be further categorised into transient and static errors (Mavis and Eaton 2002).

Transient soft errors are "*soft errors that can be corrected by repeated reading without rewriting and without the removal of power*" ("JEDEC JESD89-3A," 2007). On the other hand, static soft errors, or firm errors, are those that cannot be corrected by repeated reading but can be corrected by rewriting without the removal of power, resulting in a completely functional memory (Caywood and Prickett 1983).

When a soft error has occurred, it could result in a detected recoverable error (DRE), detected unrecoverable error (DUE) or silent data corruption (SDC) (Kadayif et al. 2010; Weaver et al. 2004). If fault tolerant techniques are implemented a soft error could potentially be recovered, either by hardware or software. This is a DRE, a more benign type of error, since recovery of the normal operation is possible.

DUE take place when the same fault tolerant techniques are able to discover and/or report an error, from which recovery is not possible. A SDC take place when an error is undetected and causes data corruption (SDC) (Constantinescu et al. 2008). In this case, the corrupted data could go unnoticed making this type of error benign, or could result in a visible error and/or catastrophic failure such as crashing a computer system.

Hard errors, or Permanent errors, lead to loss of device functionality but, in contrast with transient soft and firm errors, the functionality of the device is permanently damaged. Repeated reading, writing or repowering is not effective in recovering from this type of errors. In general, hard error effects can only be corrected via maintenance action, involving replacement of components.

A further categorisation between hard and soft errors is pseudo hard errors, sometimes referred to as power cycle soft errors (PCSE) ("JEDEC JESD89-3A," 2007). These take place as a result of the ionising radiation from a particle strike, when the functionality of the device is lost but the device is not permanently damaged. Unlike soft errors, pseudo hard errors cannot be corrected by repetitive readings or writings. Instead, they can be corrected by removing the power from the device. Examples of this are non-destructive latchup and firm errors in FPGA where the area affected by the particle strike is the control path (Edwards et al. 2004).

Although the data may not be corrupted, the device functionality is compromised. SRAM based FPGA devices are subject to this type of error if the "gate array" configuration in SRAM is corrupted. These systems contain the "gate array" configuration area within ROM, which is loaded into the SRAM during power up. Recovery can be achieved via repowering and reinitialisation.

A classification of SEEs is presented in Table 4.3. The numerous types of SEE can be categorised depending on the type of degradation, recoverability and technologies susceptibility. Long- and short-term radiation effects on different manufacturing technologies are presented in Table 4.4.

4.5.2 System Level Response

Many different acronyms are used to describe the numerous SEEs in digital integrated circuits. Also called "reversible errors", non-destructive effects can be classified as SET, SEU, MBU, MCU and SEFI.

4.5.2.1 Single Event Upsets (SEUs): Conventional Upset Mechanisms

SEUs are a particular type of SEE that take place when a single energetic particle strike causes a charge disturbance, large enough to directly modify the logic state of a sequential element, such as a register, latch, flip-flop or a memory cell. It is by far the most common effect affecting all kinds of memory devices, including SRAM, DRAM, FLASH memories, microprocessor registers, DSPs, FPGAs, logic programmable state machines and other similar devices.

SEUs can be categorised as static soft errors since the device functionality is not permanently affected (soft), and cannot be corrected by repetitive reading (static) but only through the rewriting of new data (Baumann 2005a, b; JEDEC JESD89-3A 2007).

Table 4.3 Classification of single event effects

Acronym	Name	Type of error	Affected technology
SET	Single event transients	Transient soft	Combinatorial logic, operational amplifiers, analogic and mixed signal circuits
SEU	Single event upset	Static soft	RAM, PLC—sequential logic
SBU	Single bit upset	Static soft	RAM, PLC—sequential logic
MCU	Multiple cell upset	Static soft	RAM, PLC—sequential logic
MBU	Multiple bit upset	Static soft	RAM, PLC—sequential logic
SEL	Single event latchup (microlatchups)	Pseudo-hard	CMOS, CPUs, PLC
SEFI	Single event functional interrupts	Pseudo-hard	Complex devices with built-in state or control sections
Logic SEFI	Address error, recoverable bust error, temporary block error	Pseudo-hard	Complex devices with built-in state or control sections
Soft SEFI	Resettable single event functional interrupt	Static soft	Complex devices with built-in state or control sections
Hard SEFI	Reboot or permanent single event functional interrupt	Pseudo-hard	Complex devices with built-in state or control sections
SEL	Single event latchup	Hard	CMOS, BiCMOS
Destructive SEL	Address error, recoverable bust error, temporary block error	Hard	CMOS, BiCMOS
Non-destructive SEL	Resettable single event functional interrupt	Pseudo-hard	CMOS, BiCMOS
Micro-latchup	Reboot or Permanent single event functional interrupt	Pseudo-hard	CMOS, BiCMOS
SEHE or SHE or SEHR	Single event hard error	Hard	Memories and latches in logic devices
SESB or SES	Single event snapback	Pseudo-hard	Power MOS, SOI
SEBO or SEB	Single event burnout	Hard	Power MOS and bipolar
SEGR	Single event gate rupture	Hard	Power MOSFETS, Flash memory
SEDR	Single event dielectric rupture or micro-damages	Hard	Non-volatile nMOS structures, FPGA (antifuse), linear devices

Between 1954 and 1957, there were reports of anomalies in electronic equipment during above ground nuclear bomb tests. Since these anomalies were random, and not related to any permanent hardware fault, these were attributed to electronic noise from the bomb's electromagnetic shock wave.

Even though the actual term "Single event upset" was first adopted in 1979 (Guenzer et al. 1979), SEUs were, in fact, predicted in 1962 (Wallmark and Marcus 1962)

Table 4.4 Long and short term radiation effects on different manufacturing technologies—X_1 except SOI of single event effects

Technology	Function	SET	SEU	SEFI	SEHE	SEL	SESB	SEBO	SEGR	SEDR	TID	DDD
CMOS, SOI	SRAM		x		x	X_1	x				x	
	DRAM/SDRAM		x	x	x	X_1	x				x	
	EEPROM/Flash EEPROM	x	x	x		X_1			x	x	x	
	Mcontroller/μP	x	x	x	x	X_1					x	
	FPGA	x	x	x		X_1	x		x	x	x	
Power MOS							x	x		x		
Bipolar	x	x					x	x		x	x	

when it was forecasted that terrestrial cosmic rays would lead to the eventual occurrence of upsets in microelectronics. Moreover, it was anticipated that this kind of upset would limit the volume of semiconductor devices to a minimum of about 10 μm per side.

Evidence of a small rate of cosmic ray induced upsets in bipolar J-K flip-flops in the space environment (Binder et al. 1975) was presented in 1975 confirming the earlier predictions. Four anomalies were found in the analysis of 17 years of satellite operation. It was suggested that 100 MeV heavy ions in the solar wind striking the electronics might be responsible. During the early years of computing there have been many reported cases of electronic anomalies, whose source was unknown at the time.

As an example, in 1976, the Cray1 supercomputer at Los Alamos presented an average of 25 memory parity soft errors per month. It was not until 2010 that a study was published, attributing the cause of these anomalies to high-energy neutrons from the cosmic ray background (Normand et al. 2010).

As integration density of DRAM increased to 64K, a significant SEU rate, mainly caused by alpha particle contaminants in package materials was found in terrestrial environments. The first evidence of SEUs at sea level in computer electronics was reported by May and Woods from Intel Corporation in 1978. Eventually, May and Woods attributed the anomalies to alpha particle from impurities in the packaging modules (May and Woods 1979).

SEUs at sea level and aircraft altitudes due to cosmic radiation were first predicted in 1979 by Ziegler and Lanford from IBM Corporation (Ziegler and Lanford 1979). In 1984 SEU appearances due to cosmic radiation were reported for the first time (Ziegler and Puchner 2004). The use of low alpha activity materials (May 1979) mitigated the soft error rate due to this radiation from impurities, leaving cosmic ray as the primary factor of "single event rate" (SER) (Pickel and Blandford 1978), which is the amount of single events pet unit of time.

However, the increased use of large-scale integration (LSI) technology decreased the volume of the sensitive elements, which implied a corresponding reduction of the critical charge and the number of ion pairs needed to induce a soft error. The resultant SER raise was attributed to a new source, protons from solar events and trapped protons in the Van Allen belts (Wyatt et al. 1979).

The 1980s where characterised by extensive research and development of SEU hardened electronics (Desko et al. 1990; Rockett 1988; Weaver et al. 1987) and research on the fundamental SEU mechanisms, mostly on memory circuitry (Adams and Gelman 1984; Blake and Mandel 1986), since SEUs in combinational logic were rare (May et al. 1984). In 1984 SEUs induced by atmospheric neutrons were predicted in avionics for the first time (Silberberg et al. 1984).

During the 1990s, the prediction of atmospheric neutron induced SEU in avionics was rigorously demonstrated to occur during flight (Taber and Normand 1992). Furthermore, the concern for SEU increased due to manufacturers reducing the number of SEU hardened components which led to an increased interest for commercially available off-the-shelf (COTS) components, even in space environments (Shirvani and McCluskey 1998; Underwood 1998).

Due to its high operating voltages, early SRAM cells were very robust, but with technology scaling, in the last decades, SEUs have become more of a concern, posing a major challenge for the design of memories. SEU susceptibility increases exponentially as voltage decreases and, in contrast, decreases at power 4 as feature size decreases.

Measurements of neutron accelerated induced upsets in 0.25 μm, 0.18 μm, 0.13 μm and 90 nm SRAM showed a SER/bit increase of 8 % per generation. The SER of a 90 nm SRAM increased of a by 18 % for a 10 % reduction in voltage (Hazucha et al. 2003).

In contrast, more recent results in technology nodes ranging from 250 nm through 28 nm have shown that the SEU rate per bit has been declining up to the 65 nm node (Dixit and Wood 2011). However, this long term trend has been reversed with results for 40 nm SRAM presenting 30 % higher bit SER than the previous 65 nm technology (Dixit and Wood 2011). Note that the results provided are based on bit SER. Nonetheless, for every generation the complexity and the number of bits per unit area are increasing and so is the System SER. Recent predictions using Monte-Carlo simulator CORIMS on neutron induced soft errors in SRAMS show that system SER will increase ×7 from 130 nm to 22 nm technology (Ibe et al. 2010).

Embedded DRAM has been widely used in System on Chip (SOC) systems thanks to its density and high performance. At the same technology node, the size of an embedded DRAM bit cell is a quarter of the size of an embedded SRAM cell. With scaling, the voltage reduction has also reduced Qcrit.

However, by replacing 2D capacitors (very efficient at collecting radiation charge due to its high area junctions) for 3D capacitors, the collection efficiency has decreased considerably, hence increasing Qcrit. The Qcrit increase due to junction volume scaling is more significant than the Qcrit decrease due to voltage scaling. Because of these, the DRAM bit SER has decreased to around 4× to 5× per generation (Baumann 2005a, b). Then again, the DRAM system SER has remained roughly constant over many generations.

In contrast with SRAM, whose SEU susceptibility has increased over the years, the problematic earlier DRAM based on planar cells has evolved to become one of the most robust devices.

Cell Upsets

A cell upset takes place if the deposited charge is greater or equal than the critical charge of the cell, changing its original logical value. These could be single bit upsets (SBUs), multi cell upsets (MCUs) or multiple bit upsets (MBUs).

Single bit upsets (SBUs) are single upsets in a memory cell caused by a single event, i.e. one event producing a single bit error, and are very common on SRAMs.

A single particle can energise two or more memory cells, as shown by (Reed et al. 1997). Multi cell upsets (MCUs), first reported in SRAMs exposed to the harsh space radiation environment (Blake and Mandel 1986), are multiple bit upsets for one event regardless of the location of the multiple bits, i.e. an FPGA where one routing bit gets an impact from a high energetic particle affecting several memory positions.

Hence, MCUs involve both types of upsets, the ones that can be corrected by EDC/ECC and those that cannot be. Traditionally, MCUs have represented a small fraction of the total number of observed SEU (0–5 %) (Maiz et al. 2003). However, in the case of FPGA, high linear energy transfer (LET) heavy ion induced radiation experiments indicate that as geometries shrink the MCU probability significantly increases, accounting for up to 35 % of the upsets induced (Quinn et al. 2005).

As for SRAM devices, it has been predicted that: (1) the MCU ratio will increase ×7 from 130 nm down to 22 nm; (2) the MCU maximum size (MxN bits rectangular area including failed bits) will exceed as many as 1 Mbits in the extreme case; and (3) for 22 nm process the maximum bit multiplicity will exceed as many as 100 bit (Ibe et al. 2010).

Multiple bit upsets (MBUs) also referred to as single word multiple bit upset (SMUs) (Koga et al. 1993a, b) are a subset of MCUs. And MBU is a multiple bit upset for one event that affects several bits in the same word. This type of deviation cannot be corrected by EDC/ECC. However, it is possible to partially avoid MBUs by using specific layout design of memory cells.

In contrast to cells, bit line upsets are only upset susceptible during a short period of time, the pre-charge period specific from read cycle states. However, susceptibility is dependent on the core cycle frequency. Therefore, bit line upset rates are becoming more important (Schindlbeck 2005) since recent technologies make use of shorter core cycles, which in turn involve higher susceptibility to upset.

Figure 4.11 shows the sensitive areas that are susceptible to cell and bit line upset. NMOS drains of transistors connected to capacitors are sensitive zones to cell upset.

In contrast, the sensitive zones to bit line upsets are the NMOS drains of transistors connected to bit lines (Bougerol et al. 2008).

Historically, the occurrence of MCU was attributed to the collection of charges generated by a nuclear spallation reaction as a result of the impact between a secondary ion and the device. As sensitive devices shrink, neighbouring cells present closer physical proximity, increasing the number of cells that can be affected by the impact of a single particle. Nonetheless, novel MCUs are being reported such as "charge sharing among neighbour nodes" (Amusan et al. 2006; Ibe et al. 2006).

4.5.2.2 Single Event Transient (SET): An Emerging Upset Mechanisms

Without the peripheral logic that interconnects them, sequential logic including embedded SRAM and DRAM would be useless. In general, the scientific community is mostly concerned with the effects of SEUs on sequential logic even though combinational logic is not immune to radiation as single event transients do occur here as well (Baumann 2002; Buchner et al. 1997; Zhu et al. 2005).

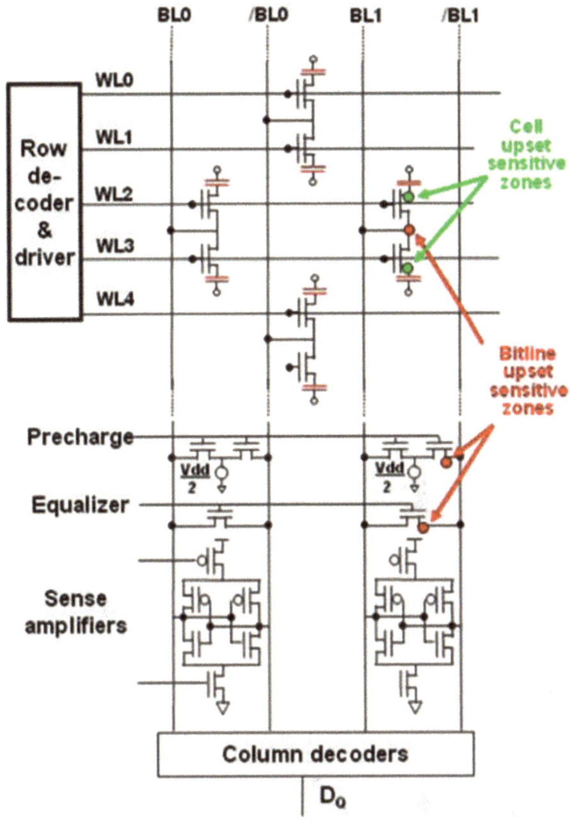

Fig. 4.11 Sensitive areas to SEU in a DRAM memory array (Bougerol et al. 2008)

However, confusion seems to exist in the literature regarding the terminology used for single event transients. In analogue circuits, a SET has also been referred to as "analogue single event upset" (Ecoffet et al. 1994). In digital circuits, a transient that causes an incorrect state in the data output of a logic gate has been referred to as "digital single event upset" (Reed et al. 1996).

Earlier publications often incorporate both phenomena, SET and SEU, together as SEU, perhaps because the effects of an SET can potentially be propagated down the logic line and change the state of a sequential logic element. In this case, the effects are identical to the effects produced by an SEU as shown in Fig. 4.12. It is also possible that more than on logic element change their state. This is known as a *single event multiple upset* or SEMU and should not be confused with MBU/MCU.

In contrast with SEUs, SETs were at the time not considered a serious threat to the reliability of semiconductors.

For the purposes of our work, the following definition will apply to the term SET: Single Event Transients (SETs) are analogue transient pulses resulting from a single ionising particle, that are large or big enough to momentarily change the output of non latched elements, such as combinational logic, clock line and global

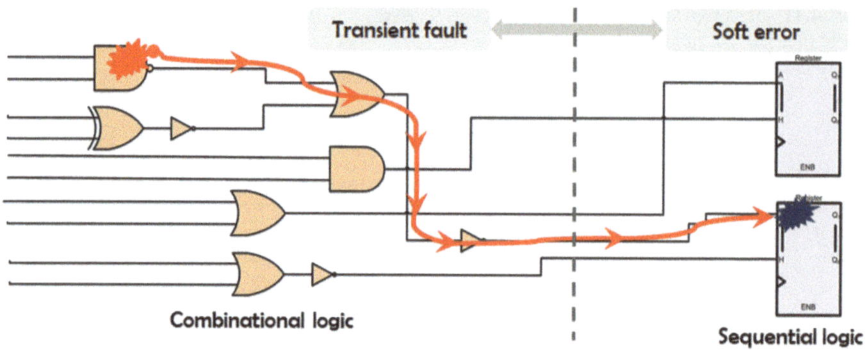

Fig. 4.12 Traditional propagation of an SET in combinational logic

control lines to an incorrect logic value. The duration of such pulse is in the order of 100_{ps} (Pouponnot 2005).

As seen previously in Sect. 4.5.1.2, different semiconductor technologies show different charge collection and transport mechanisms that lead to different pulses. Depending on the device technology, circuit topology, impact location, particle energy device supply voltage and output load, the resultant SET would have unique characteristics in terms of amplitude, waveform, polarity, duration, etc. Pulses can vary from tenths of picoseconds to tenths of microseconds.

The effects of a SET can further be propagated along the logical path, and potentially be latched into one or more flip-flop, latch or register at a distant location from the original charge collection area. Yet, there has not been too much interest in protecting combinational logic since this type of logic has a natural tendency to mask these transient faults.

There are inherent masking mechanisms that mitigate the propagation of the glitches, preventing the latch from taking place. These three mechanisms, that can provide a certain level of natural resistance to soft errors, are logical masking, electrical masking and latch-window masking (Shivakumar et al. 2002; Wirth et al. 2008).

Logical masking takes place when the particle strikes a portion of the combinational logic that, regardless of its output, has no effect on the output of the subsequent gate, Fig. 4.13.

The result of the subsequent gate is solely determined by its other input values. For instance, the output of a NAND gate with an input A equals to '1' and an input B equals to '0' would not be affected by a glitch on the A input since regardless of the value that A has, the gate's output would be '1'.

Electrical masking occurs when, as the signal propagates, due to the electrical properties of the subsequent logic gates, the pulse suffers from attenuation to a point that it is not of sufficient magnitude to upset any downstream state element (Fig. 4.13).

Latch window masking, also called timing windows masking, occurs when the undesired pulse reaches a latch at the wrong time of the clock transition

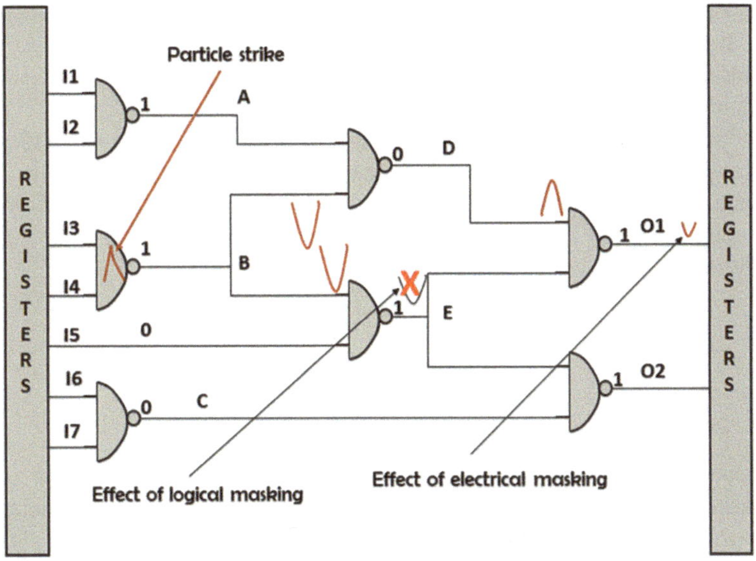

Fig. 4.13 Effects of logical and electrical masking on a pipeline stage (Ramanarayanan et al. 2009)

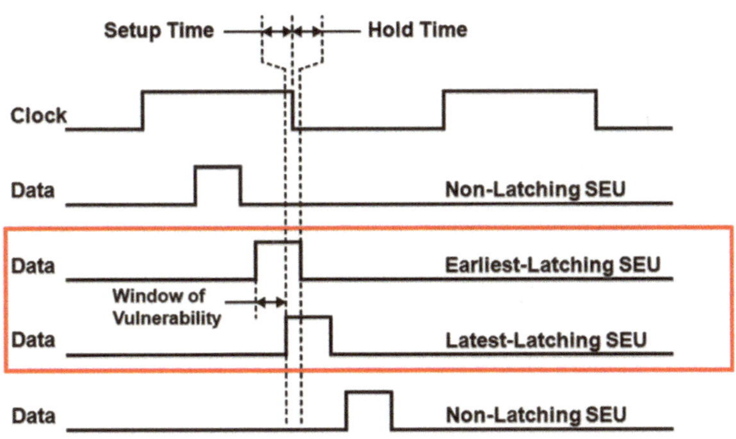

Fig. 4.14 Latch window masking; temporal relationship of latching a data SET as an error (Mavis and Eaton 2002)

(Cha et al. 1993). In other words, the pulse does not satisfy the compulsory set-up and hold time of the flip-flop. The transient will get latched if the pulse reaches the latch within the "window of vulnerability" (Fig. 4.14), hence causing data corruption.

In terms of upset tolerance of single gates, there are two characteristics of interest: glitch generation and glitch propagation (Dhillon et al. 2005). The shape and the magnitude of the voltage glitch generated at the gate's output are determined by the glitch generation characteristics. The voltage magnitude of the glitch depends on the total capacitance of the node, while the duration of the glitch depends on the gate's delay. Faster gates lead to wider glitches and therefore better generation characteristics.

Alternatively, the glitch propagation characteristics of a logic gate determine the glitch attenuation as it passes through the gate. Assuming a linear ramp at the output of a gate, where d is the gate propagation delay and w_i is the glitch duration at the gate input, the glitch duration of the gate w_0 can be approximated using Eq. (4.4) (Dhillon et al. 2005) as:

$$\begin{aligned} w_0 &= 0 \quad \text{if} \quad d > w_i \\ w_0 &= 2(w_i - d) \quad \text{if} \quad 2d > w_i > d \\ w_0 &= w_i \quad \text{if} \quad w_i > 2d \end{aligned} \tag{4.4}$$

According to Eq. (4.4), slower gates will induce more attenuation on glitches than faster gates. Therefore, fast gates have better glitch propagation characteristics. An increase in the gates capacitance would increase the delay of the gate, which in turn, would reduce the glitch propagation characteristics. SETs affecting the clock logic or the reset trees can lead to much larger problems (see Sect. 4.5.2.3).

In the past, these masking effects are some of the reasons why SETs have not been a dominant contributor in the overall SER. In addition, designers have not been significantly concerned about errors in microprocessor logic because the number of flops on microprocessors was much fewer than the number of memory cells. Since flop protection techniques are more difficult to implement than memory protection mechanisms such as parity or ECC, from 90 nm downwards, flop SEU rates are higher than SRAM SEU rates.

SETs are particularly worrisome in safety-critical applications whose memory has been protected to decrease SEU rates. In this type of systems, SET rates can be the dominant reliability failure mechanism.

4.5.2.3 Single Event Functional Interrupt (SEFI)

SEFI represent the most disruptive version of non-destructive SEE. Although this type of anomaly was previously predicted for space environments (Koga et al. 1985), the term single event functional interrupt (SEFI) was first mentioned in 1996 (EIA/JEDEC STANDARD 1996).

SEFI is defined as all non-destructive failure modes that lead to the malfunction (or interruption of normal operation) of a part or the totality of the device (Bougerol et al. 2008). This definition is in contrast with certain authors that define SEFI as the cause of a higher error rate than expected due to uniformly distributed upsets (Crain et al. 1999; LaBel et al. 1996).

Table 4.5 Classification of SEFI

Name	Also called	Typical effect	Recovery procedure	Technology affected	Examples
Logic SEFI	Address error, recoverable bust error, temporary block error	Reading/writing of the wrong row, column; 512-8K addresses in errors	Rewriting of the right value	Complex memories such SDRAM	Fuse latch upsets (SEFLUs)
Soft SEFI	Resettable SEFI	Functionality loss of up to a full memory bank	Refresh cycles	FPGA, microprocessors, complex memories	Stuck block errors
Hard SEFI	Permanent SEFI, Reboot SEFI	Complete loss of functionality	Complete power cycle of the device	FPGA, microprocessors, complex memories	Events that induce data and functionality loss that cannot be recovered

The causes and effects of SEFIs vary from the type of component and the technology used. In general, SEFIs are linked to an upset (SET or SEU) in a control area that configures a specific function, and leads to the loss of that function. In contrast to SEUs and SETs that may or may not affect the operation of the device, every single type of SEFI leads to a direct malfunction.

Figure 4.12 shows an SET affecting combinational logic, not affected by the logical and electrical masking mechanisms (as in Fig. 4.13), that propagates to a register in a control area within the latch window (as in Fig. 4.14). If the register affected is being used by a vital part of the system software, a SEFI could take place.

As microcircuits become more complex they also become more susceptible to SEFIs; among those: SDRAMs (Harboe-Sorensen et al. 2007) with complex internal architecture (such as state machine), FLASH memories (Irom and Nguyen 2007; Nguyen et al. 1999; Oldham et al. 2008), FPGA (Czajkowski et al. 2006) and microprocessors (Czajkowski et al. 2005). Dependent on cause, consequences and recovery procedures, SEFIs can be classified as logic, soft or hard (see Table 4.5):

Logic SEFIs (Bougerol et al. 2008): with regard to memories, it is also called "address error", "recoverable burst error" (Ladbury et al. 2006) or "temporary block error" and mainly includes row and column errors. The upset of a row or column register leads to the reading or writing of the wrong row/column. This type of SEFI typically causes between X and 8 k addresses in errors where X is the number of addresses per row/column (Bougerol et al. 2008). Rewriting of the right values is used as to recover functionality (Schagaev and Buhanova 2001).

Examples of logic SEFIs are "fuse latch upsets" also called SEFLUs (Bougerol et al. 2010, 2011) that lead to the wrong addressing of a whole row/column. Manufacturers are experiencing an increasing number of defective cells, therefore adding spare cells and exposing them to reliability tests. If during those tests, a cell is found defective, fuse latches are used to disable the particular row/column. Typical signatures of fuse latch upsets are multiples of X addresses where X is the number of addresses belonging to a column/row.

Soft SEFIs also called “Resettable SEFIs” (Bougerol et al. 2008; Lawrence 2007) are due to upsets in the device configuration area and usually induce the functionality loss of several thousands of addresses up to a full memory bank. Reconfiguration of the device with a mode register set command can be used as a recovery procedure of the functionality (but not the data). Examples of this are “block SEFIs” also called “stuck block errors”, observed in the IBM Luna-ES rev C during heavy ion testing (“NASNGSFC Landsat-7 Project Office, Private Communication,” 1995) where an entire row of 1,024 addresses was stuck to a specific value. Since simple writing was not sufficient, device refresh cycles were used to clear the problem. SEUs in selected areas of an FPGA such the JTAG bit serial configuration port can lead to inability of reconfiguration.

Hard SEFIs (Bougerol et al. 2010; Harboe-Sorensen et al. 2007), also called Reboot SEFIs (Bougerol et al. 2008), “permanent SEFIs” (Slayman 2005), “non resettable errors” (Lawrence 2007, p. 512) or “persistent non recoverable errors” (Ladbury et al. 2006) can be induced by different phenomena and lead to the complete loss of memory functionality. Possible causes of this type of catastrophic SEFI are upsets in the internal state machine, counter registers or activation of special modes. An example of this is an SEU in one of the power on reset registers that can lead to the removal of the entire configuration area. Complete power cycle of the device is compulsory as a recovery procedure.

Fortunately, the probability of SEFI is low compared to other types of SEEs (Slayman 2005). The reasons for that are:

1. The ratio of the periphery logic area to memory array area is very low;
2. The critical charge for logic gates is usually higher than for SRAM cells.
3. The most part of the periphery logic is combinational, and therefore less susceptible to upsets due to the three inherent masking mechanisms.

SEFIs can also be classified as high current SEFIs if they involve a certain increase in current (Koga et al. 2001a, b).

In addition to SEFIs in complex memories, the energetic particles can also strike other circuits such that the error detection and correction mechanisms affect the functioning of the whole circuit. In FPGAs, SEFIs can cause the device to stop from functioning normally and therefore require a power reset in order to resume normal operations.

In microprocessors, SEFIs can induce upsets in the program counter, illegal branching and jumps to undefined states.

4.5.2.4 Single Event Latchup (SEL) and Other Destructive Effects

Also called “hard errors” or “non reversible errors”, “single event destructive effects” are events that momentarily or permanently change the state of a device or cell/node affecting their functionality. Destructive effects are persistent even after a reset or reconfiguration and a replacement of components may be required.

Single Event Latchup

A latchup is an unintended and potentially catastrophic state that affects CMOS devices, characterised by excessive current flow between a power supply and its ground rail.

It can take place due to the interaction between parasitic structures, usually an npn- and a pnp-bipolar transistor.

A low resistance path develops between ground and power supply of the device and remains after the triggering event has been removed. Once triggered, a latchup can amplify currents to a point where the device fails as a result of thermal overstress. This electrically induced effect typically occurs in improperly design circuits.

However, it was demonstrated (Leavy and Poll 1969) that a latchup can also be induced via ionising radiation (SEL), including high-energy protons, alpha particles, cosmic rays and heavy ions. The difference between a conventional latchup (electrical) and a single event latchup (SEL) is that latter phenomenon is triggered by an energetic particle instead of an electrical overvoltage. A classification of different SEL is shown in Table 4.6.

Parasitic transistors of CMOS devices can be triggered by the strike of high-energy protons, alpha particles, neutrons and heavy ions. An SEL may occur if enough energy, critical charge, is deposited by a given particle within a microscopic region of the device, regardless of the total flux. High currents can lead to metal traces to vaporise, bond wires to fuse open and silicon regions can be melted due to thermal runaway. Hence, the latched condition may potentially destroy the device, affect other surrounding devices and destroy the power supply (traditional or destructive SEL).

Both high current and low current SELs can occur (LaBel et al. 1992). Modern devices may have many different latchup paths, making characterisation of those latchup states a challenging task. In some cases, events resulting in localised high current (micro-latchups) can remain functional.

Table 4.6 Classification of SEL

Name	Type of error	Nature	Recovery procedure
Traditional or destructive SEL	Hard	High current	Replacement of components
Non-destructive SEL	Pseudo-hard	Low current	System restart
Micro-latchup	Pseudo-hard	Localised, high current	Reducing the power supply voltage below the holding voltage or reset

In order to restore the device to a normal operation, these effects can be tolerated by reducing the power supply voltage below the holding voltage e.g. power off-on reset (PCSE). An example of this phenomenon is the latchup susceptibility of the Pathfinder's modem used in the Mars mission (Matijevic 1996). The latched states with periods of up to 1 h were not destructive (micro-latchups), and thus, the modem was adapted with additional circuitry and software to detect the event and then correct it by powercycling the device.

Additionally, latent damages have been observed in several types of CMOS devices after non-destructive latchup events (Becker et al. 2002). Becker defines latent damages as "*structural damages that cause no electrically observable parametric or catastrophic device failure, but can be detected by surface analysis using optical or scanning electron microscopy*". These type of permanent structural damages are a potential reliability hazard since the interconnect cross-sections in the damaged area may be reduced by one or two orders of magnitude.

Sometimes the SELs are not localised and affect the entire device, but the current may not be high enough to destroy the device (non-destructive SEL). Therefore, SELs are not invariably destructive and can also be categorised as pseudo hard errors.

Temperature is an important factor in SEL susceptibility. Higher temperatures involve a cross section increment and reduction of SEL threshold (Johnston et al. 1991).

SELs can be mitigated through internal fabrication process modification. Silicon on insulator (SOI), silicon on sapphire (SOS) and the use of epitaxial substrates are immune to this type of effects (Miller and Mullin 1991). However, those are very expensive and their availability normally limited to mission critical systems in space environments (Pouponnot 2005).

Additionally, different layout techniques, like guard drains and guard rings, are often used in CMOS processes. Alternatively, SEL can be circumvented externally through the use of current sensing, watchdogs, etc. Internal methods are trying to keep the event from occurring. With external mechanisms, the event still occurs, but there should be a recovery strategy to deal with them.

Single Event Hard Error (SHE or SEHR) or Stuck Bits

Since the mid-1980s certain SRAM devices, when exposed to heavy ions, experienced semi-permanent stored bit patterns or stuck bits with no implication of total dose effects. This form of damage was not reported until 1991 (Koga et al. 1991) and was later studied and renamed as "single hard error" (SHE) (Dufour et al. 1992).

SHE is an unalterable change of state of a memory element associated with semi-permanent damage due to high-localised dose deposition from a single ion track. This type of effect affects memories (SRAM, DRAM, Flash) and latches in logic devices rendering the cell un-programmable (Dufour et al. 1992).

The cell may have an indeterminate value, also appearing as a permanent fault at the system level. SHEs are considered semi-permanent since some of the stuck bits tend to disappear (in some cases after a day (Duzellier et al. 1993)).

Single Event Snapback (SES or SESB)

This type of effect induces high currents in most cases and is particularly difficult to differentiate from high current SELs (Beitman 1988; Koga and Kolasinski 1989). While SESBs can take place in technologies immune to SEL, it does not require a four-region structure to arise. In this context, snapback has been confirmed to be particularly susceptible to SOI structures because of their internal design (Dodd et al. 2000). With regard to SESB and NMOS technology, the parasitic NPN bipolar transistor that exists between the drain and the source amplifies the avalanche current resulting from the impact of an ionising particle. The transistor then opens and remains open.

Like SEL, SESB is also considered a potentially catastrophic event since it can lead to device destruction if not corrected within a short time of occurrence. The main differences between SEL and SESB lie in the amplitude of the current increase, their temperature dependence and recovery conditions. First, unlike destructive SEL, it is often possible to restore normal operation and bring the device out of the high current mode by changing the gate voltage without shutting off the power supply. Secondly, the amplitude of the current increase is much lower for SESB due to its localised nature.

Finally, contrary to SEL, SESB is weakly dependent on temperature (Johnston 1996). These facts can be used to distinguish between SESB and SEL mechanisms.

Single Event Burnout (SEB or SEBO)

SEBO typically occurs in power metal oxide semiconductor field-effect transistors (power MOSFETs) and bipolar transistors since these devices contain parasitic bipolar transistors between the drain and the source (Hohl and Galloway 1987; Waskiewicz et al. 1986). SEBO creates a permanent short between a source and a drain and involves high currents and localised overheating.

If the device is not provided with current limitation capabilities, and the drain-source voltage exceeds the local breakdown voltage of the transistor, the SEBO can lead to the destruction of the device by melting of the silicon in the affected region (Stassinopoulos et al. 1992), illustrated by see Fig. 4.15.

It has been shown (Johnson et al. 1992) that higher temperatures decrease the SEBO susceptibility. The probability of SEBO occurrence is low, but apart from the selection of immune device technology, there are no mitigation techniques.

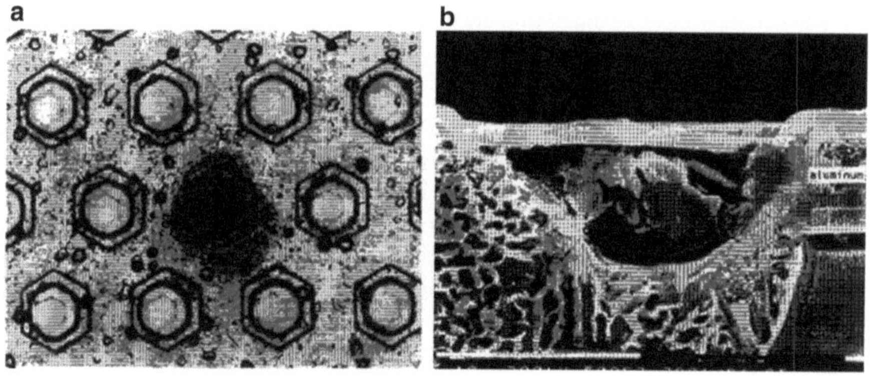

Fig. 4.15 IRF 150 power MOSFET burnout: (**a**) Optical view of burnout area on the surface, (**b**) Scanning electron microscope (SEM) sectional view of a burnout area with ×1,000 magnification (Stassinopoulos et al. 1992)

Single Event Gate Rupture (SEGR)

It was first observed in non-volatile memories in 1980 (Pickel and Blandford 1980) and later identified and confirmed in 1984 (Blandford et al. 1984). In 1987 was reported in power MOSFETs (Fischer 1987) but due to the scaling of CMOS technology SEGR has become a concern in low voltage circuits (Silvestri et al. 2009).

This type of single event is often observed with SEB in power MOSFETs. SEGR is triggered by a single ionising particle in a high field region of a gate oxide, creating a localised gate rupture in such area (Sexton et al. 1997). This rupture manifests as a permanent conducting path between the gate and the drain (gate rupture—see Fig. 4.16). As a result, the electrical performance is compromised and the functionality of the device may be affected.

Flash memories (Oldham et al. 2006) and non-volatile SRAM are SEGR susceptible during a write or clear operation due to the large voltage applied to the memory elements.

Like SEBO, the probability of occurrence is low, but should be taken into account in the component selection process. In order to mitigate SEGR, voltage derating and limiting the available energy to a device can be employed.

Single Event Dielectric Rupture (SEDR)

Also called "micro damages", SEDR was encountered during heavy ion SEE testing of antifuse FPGA (Katz et al. 1994) and eventually identified as ion induced rupture of antifuses. Similar to the SEGRs observed in power MOSFETs, SEDRs affect non-volatile NMOS devices and non-volatile FPGAs (Katz et al. 1997; Swift and Katz 1996). SEDRs are triggered by a single ionising particle, and lead to the formation of a conducting path in a high field region of a dielectric.

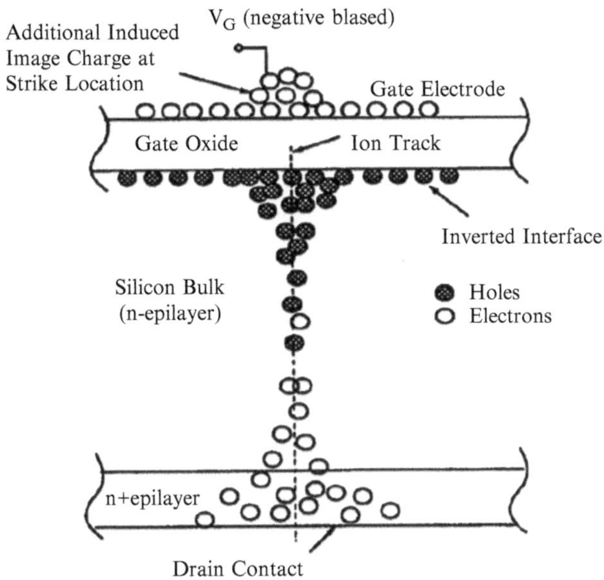

Fig. 4.16 SEGR as a result of the impact of a highly energetic particle. Holes from the particle's track aggregate under the gate oxide increasing the high field of the gate oxide to the dielectric breakdown point (Allenspach et al. 1994)

4.6 Conclusion

- Radiation can have a major impact on all kinds of embedded microelectronics potentially leading to catastrophic failures.
- As we move to denser semiconductor technologies at lower voltages, system single event rate will continue to rise and in particular the contribution of single event upsets, single event transients, multi-cell upsets and single event functional interrupts will increase.
- Error correcting codes are not efficient when dealing with certain multi-bit faults and errors in combination logic. In the case of safety-critical embedded systems, more efforts need to be directed towards research on mitigation techniques for the recent and future undesired effects.
- This chapter presents an analysis of the long-term cumulative and short-term effects of radiation on embedded systems. First, we make an overview of the fundamental damage mechanisms and, resulting from such mechanisms we introduce the major macro effects. Second, we focus on the short-term degradation induced by ionising particles, namely single event effects. Third, we describe the physical mechanisms that are responsible for SEE including charge deposition, charge transport, charge collection, to finally fully describe the different circuit responses.
- As a result, an extensive taxonomy of single event effects has been produced, describing their nature, type of degradation, susceptibility, fault rate trends and recoverability.

Chapter 5
FT Models

5.1 Models

We define M as the known model of a system that performs a given function F. Let us imagine a new feature of extreme reliability in that model. In order to express the existence of this new feature, the predicates P and Q are introduced to determine the state of the model. P and Q also defined the direction of the time arrow (see Fig. 5.1).

To analyse methods for achieving a required level of reliability with performance and power consumption constraints, we offer a combination of the following three models:

- The model of the system M_s
- The model of the faults M_{fault} that a *RT FT* system will be exposed to
- The model of fault tolerance M_{FT} or the new structure that implements *FT*

As shown in Fig. 5.1, M_s, M_{fault} and M_{FT} are mutually dependent models. Notice that in this approach development and manufacturing costs of a solution are not considered.

M_{fault} is a description of all faults that a system must tolerate. In binary logic a typical permanent fault can manifest as "stuck at zero" or "stuck at one".

Table 5.1 shows typical examples of HW faults. Hidden faults, also called Latent faults are behavioural faults that exist in the hardware over a long period of time, e.g.: Byzantine faults[1] and fail-stop[2] faults. Both types complicate the design of *FT*;

[1] Byzantine faults occur when a faulty system continues to operate, producing incorrect results sometimes giving the impression that they are working correctly. Dealing with this type of fault is difficult.

[2] Fail-stop (also known as fail-silent) faults take place when a faulty unit stops functioning, producing no bad output. Fault stop assumes to produce no output - i.e. freezing the output, that clearly indicates that units has failed.

V. Castano, I. Schagaev, *Resilient Computer System Design*,
DOI 10.1007/978-3-319-15069-7_5

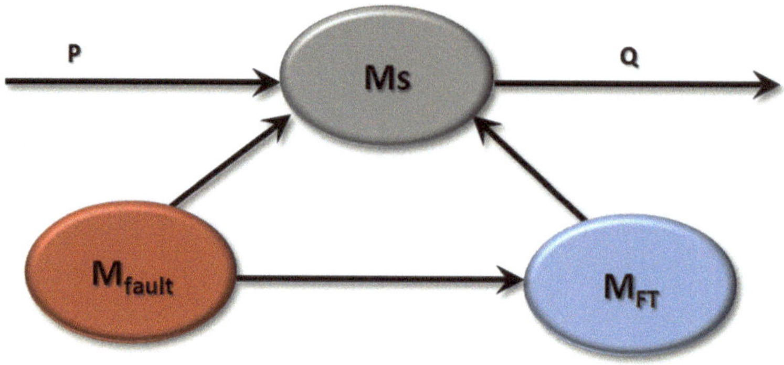

Fig. 5.1 New feature of an *FT* system: reliability

Table 5.1 Typical example of HW faults

Type of fault	Description	Impact
Byzantine	The behaviour of a component that gives conflicting values to other components	The entire system is affected
Subsystem fault	A temporary or permanent incorrect behaviour of a subsystem	The entire system is affected
Open fault	Resistance on either a line or a block due to a bad connection	The value associated to the line or the block is modified
Bridging fault	Signals S_1 and S_2 are connected unintentionally	The value associated to the line or the block is modified to a different value
Stuck-at fault	The result value is fixed to *0* or *1*	The result value is stuck to *0* or *1*
Bit-flip fault	A state switch from *0* to *1*, or vice versa, when it should not	The result changes its original value

all described faults should be tolerated within a limited and specified period of time. This period actually determines the availability of the system. Fault types differ by their impact, as well as the way they are handled.

Thus, the fault model has its own hierarchy, including single-bit, element, behavioural and subsystem faults. One has to accept that the fault type is varying and some action hierarchy to tolerate them is also required. A detailed fault model is further developed and discussed in Sects. 5.2–5.4.

Fault encapsulation approaches can help to handle faults: due to deliberate design solutions it is possible to ensure that severe faults in the system do not escalate and remain simpler to handle, therefore making the fault handling practically possible to implement.

RT FT system applications assume long operational life; however, fault-handling schemes are needed much more often towards the end of the device life cycle. The appropriate techniques for tolerating faults of various types are presented on Table 5.1. As discussed in Sect. 3.6.1 to tolerate transient faults, time redundancy in hardware (e.g. instruction re-execution) might be effectively used and implemented.

System software support is also needed, as the hardware cannot cover all possible faults.

It is obvious that faults, occurring at the bit level (stuck zero, stuck one and similar) should be efficiently handled *ASAP* (as soon as possible) and *ALAP* (as local as possible), i.e. at the same or nearest level. The term "*level*" in our case means the level in the hardware hierarchy on which the fault should be handled.

In other words, when a "*stuck-at zero*" permanent fault has occurred in the register file (RF) with no corrective schemes available, the whole RF has to be replaced, if no other possible reconfigurations were predefined. In turn, when only one RF is integrated in the chip and no other reconfigurations are defined then the whole chip has to be replaced, etc. Pursuing these two principles allows limiting the fault spreading and its impact to a higher level either in the chip or the system as a whole.

Example: To tolerate bit-flip faults, hardware and system software information redundancies might be used, as well as hardware structural support. In this sense parity checking in registers, supported and implemented concurrently by hardware, is described as *HW(δI)*. *HW(δS)* and *HW(δT)* are needed as supportive redundancies, *HW(δS)* describing the additional parity line and comparison logic, and *HW (δT)* the additional time needed to update the parity line and executing the comparison. However, the main type of redundancy used in this approach is information.

An exact characterisation of the distribution of faults for computer systems is extremely difficult due to the number of different factors that determine faults, such temperature, vibration, radiation exposure etc. Besides, discriminating between transient and permanent faults is difficult. The transient–permanent fault ratio varies from 10 to more than 1,000 depending on the technology, manufacturing scale, operating conditions, etc. In the case of memories a typical value of hard error rates is in the order of 10–100 FIT whereas for soft errors it can vary between 1,000 and more than 5,000. The upper bound belongs to aerospace and aviation, principally due to faults induced by alpha particles.

Figure 3.2 and Fig. 5.1 are transformed into Fig. 5.2 which presents various faults in the system and various possible solutions. M_{fault} illustrates the fact that the fault types are not separated. For example, Byzantine faults of the system might be "*stuck at zero*" faults of the hardware that were spread throughout the system. The latency of faults thus becomes crucial in determining the reliability of the system. Consequently different faults require different actions and mechanisms to tolerate them.

The system model of Fig. 5.2 has overlapped *SSW* and *HW* ellipses to represent the duality of the system: hardware and system software. Both of them must be involved in the implementation of fault tolerance and real-time features.

The overlapped *HW* and *SSW* ellipses indicate that HW and SSW functions might be applied to tolerate specific types of hardware faults. Other fault types might also be tolerated by HW or SSW only. M_{ft} is "a conceptual deliverer" of

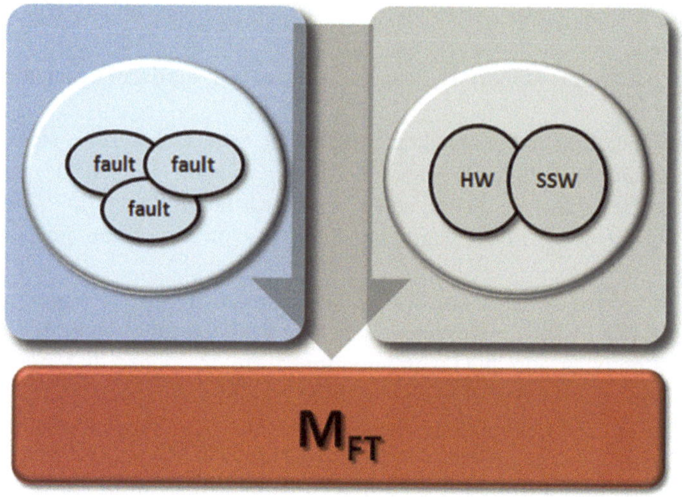

Fig. 5.2 Fault tolerance: a model of a computer system

reliability for the *RT FT* system. It has to be effective during the whole operational lifetime of the computer itself.

As hardware degrades over time, the fault tolerance mechanisms are more likely to be used towards the end of the life cycle. FT systems are designed with the assumption that new types of faults do not appear during the operational life time of the system, i.e. the system must be designed to be fault tolerant for the set of faults and their types known at design time. All these solutions require careful analysis due to their impact on the system reliability.

In contrast to the usual assumption in reliability modelling, one has to assume that a fault might exist in the system over an arbitrary long period of time (latent fault) and its detection and elimination is not possible "at once". Consequently we accept that FT is a process, and discuss it in the following sections. Using Dijkstra's approach (Dijkstra 1965) of defining a function as a process described by its algorithm, we consider FT as a function that is also described and implemented by an algorithm.

There are several options to achieve fault tolerance assuming the use of *HW* and *SSW* by using various types of redundancy mentioned above. However, the use of certain redundancy types might cause system performance degradation which is especially true for software measures (Kulkarni et al. 1987; Oh et al. 2002a). Further analysis of performance/reliability degradation should be taken into account.

The introduced system redundancy might be used in a way to tolerate only certain fault types, thus degrading fault coverage, keeping performance at acceptable levels. Software based redundancy might preserve the same type of fault coverage but with more time redundancy—delays (recovery time degrades, availability degrades), but the fault coverage might not, and thus, the system degrades in terms of reliability.

5.2 Model of Fault

It is unfeasible to describe all possible faults that may occur in a system. In order to make the evaluation of faults possible, they are assumed to behave according to some fault model. A *fault model* (*FM*) is considered as a way of summarising many fault descriptions at once (Dunn 1991). Often it is desirable to discuss many different faults at the same time and to describe their common characteristics.

Fault models are used to represent in a simple form the consequence of complex physical mechanisms that lead to faults. In the case of electronic systems, the modelling of faults can be implemented at two different levels: at the level of hardware components that implement a system (e.g. memory subsystems, register banks, *ALU*) or at the system level. The latter is directly related with the information that the system manipulates (e.g. instructions and data program).

The simplest *FM* is to consider the logic gate as a single unit with a constant failure rate, instead of considering different failure rates for the individual transistors that form the unit. As in 0, analysing the physics of faults to the atomic and molecular level would provide a clear understanding of the failure mechanisms. Such understanding is very helpful in the development of fault models. Primarily based on the work of (Avizienis et al. 2004), we extend the classification of faults depending on the way they are originated or manifested.

5.3 Classification of Faults by Origin

Faults can be classified differently depending on attributes related to their origin, including their cause, the level at which they take place, the phase of creation, nature, system boundaries, phenomenological causes and intention. Table 5.2 shows a number of faults classified by their origin attributes.

5.3.1 Level Response

Hardware defects can be the source of *physical faults*. *Logical faults* can be used to model the manifestation of physical faults on the behaviour of a system. Logical faults represent physical faults in order to simplify the logic function of the circuit and fault analysis process. They can be subdivided into *structural faults*, which are related to structural models and modify the interconnection among components, and *functional faults*, which are related to functional models and change the functions of components and circuits.

Component faults are a type of structural faults, which can be applied at the *transistor level*. Some of these are:

- *Stuck-open or stuck-off*: a transistor is always off and not controllable by input
- *Stuck-short or stuck on*: a transistor is always on and not controllable by gate input

Table 5.2 Classification of faults by origin

Level of abstraction	Structural faults	Transistor level (component faults)	Stuck-open or stuck-off
			Stuck-short or stuck-on
		Gate level (interconnect faults)	Stuck-at (Armstrong 1966; Galey et al. 1961): s-a-0, s-a-1
			Timing delay (Smith 1985): path-delay (PDF) and gate-delay (GDF) faults
			Bridging (Mei 1974)
	Functional faults	Pattern sensitive faults (PSF) (Hayes 1975)	Passive PSF
			Active PSF
			Static PSF
		Coupling faults (CF)	
Phase of creation of occurrence (Landwehr et al. 1994)	Development faults	e.g. specification faults, implementation and manufacturing faults	
	Operational faults	e.g. ageing faults e.g. alpha particle hits	
System boundaries (Avizienis et al. 2004)	Internal	e.g. design and implementation faults	
	External	e.g. radiation, temperature changes, power surges from external power supply	
Phenomenological cause	Natural (Jennings 1990)		
	Human-made (Hugue and Purtilo 2002)		
Capability/objective/intent (Brocklehurst et al. 1994)	Malicious	Deliberate	
	Non-malicious	Accidental	
		Incompetence	
Nature (Avizienis et al. 2004)	Hardware	Cell errors, combinational logic errors. . .	
	Software	Branch errors, missing instructions, missing pointers. . .	

Cause				
	Specification mistakes			
	Defects	Implementation mistakes		e.g. Pentium FDIV bug (Coe et al. 1995; Price 1995)
		Manufacturing defects		Global defects or gross area defects (Koren and Koren 1998)
				Spot defects (Koren and Singh 1990)
	Operating environment—external disturbances	Thermal stress		
		Heat		
		Electro-migration EM		
		Voltage drop		
		Noise	Electrical overstress	Hot carrier injection HCI (DAHC, CHE, SHE, SGHE)
				Negative bias temperature instability (NBTI)
				Latchup
			Induced charging	
			Oxide breakdown	
			Radiation	See Tables 4.1, 4.2, 4.3, 4.4, 4.5, and 4.6
			EMP	

Another type of structural faults, *interconnect faults*, can be applied at the *gate level*. Among these:

- *Stuck-at faults* (*SAF*): single or multiple lines have a constant value of 0 (*s-a-0 faults*) or 1 (*s-a-1 faults*) regardless of the value of the other signals in the circuit (Armstrong 1966; Galey et al. 1961)
- *Timing/delay faults*: certain defects due to manufacturing or external reasons do not change the logic function of components, but can cause timing violations; faults due to propagation delays along a path (*path-delay faults*, *PDF*) or gate (*transition* or *gate delay fault*, *GDF*) (Smith 1985), exceeding the limits required for correct operation
- *Bridging faults* (*BFs*): two or more distinct lines are shorted (Mei 1974) usually due to particles or shorted metal line. Depending on the value the bridging could be *AND bridging* (also referred to as *0-dominant*), *OR bridging* (also referred to as *1-dominant*) or *Indeterminate Bridging*. It is obvious that the probability of BF's occurring increases with (1) shorter distances between metal lines due to the use of shrinking technology and (2) the use of long parallel lines (Tehranipoor et al. 2012).

Different types of functional faults that can be applied at the *functional level* are:

- *Pattern sensitive faults* (*PSFs*): where a fault signal depends on the signal values of nearby components (Hayes 1975); typical in DRAM, there are three types of PSFs due to changes in the neighbourhood pattern:
  - Passive PSF: the value of a cell remains
  - Active PSF: the value of a cell changes
  - Static PSF: the value of a cell is being forced to a particular state (0/1)
- *Coupling faults* (CF): A subset of SPF, represent a specific pattern sensitivity between two memory cells (Nair et al. 1978); Two memory cells C_j and C_j are coupled if a transition from X to Y in one cell, say C_i, changes the state of the other cell, given that:

$$X \in \{0, 1\} \quad \text{and} \quad X = \overline{A}$$

- Idempotent *coupling faults*: a transition $0 \rightarrow 1$, or $1 \rightarrow 0$ in C_i forces the contents of C_j to a specific value $X \in \{0, 1\}$
- Inversion *coupling faults*: a transition $0 \rightarrow 1$, or $1 \rightarrow 0$ in C_i forces an inversion $0 \rightarrow 1$, or $1 \rightarrow 0$ of C_j.

5.3.2 *Cause of Faults*

5.3.2.1 Specification Mistakes

Specification mistakes, which take place during the planning and design phases, can be the source of faults (*specification faults*), including incorrect timing, power and environmental requirements. The effect of certain specification faults may be corrected via fault masking.

5.3.2.2 Defects

A hardware defect in electronics is the unintended difference between the implementation and the intended design. Implementation mistakes, such as the Pentium FDIV bug (Coe et al. 1995; Price 1995), are a type of defects. Conversely, imperfections in the fabrication process of state of the art VLSI technologies result in manufacturing defects, whose severity increases proportionally with the size and density of the chip.

Manufacturing defects are largely dependent on the specific technology and layout, and include processing and material defects such: dust particles on the chip, conducting layer defects (shorts and opens), oxide defects, scratches and gate oxide pinholes, defects caused by either extra or missing material (Koren and Koren 1998).

Manufacturing defects can be classified as *global defects* (or *gross area defects*), affecting large areas of a wafer and so can be easily detected during manufacturing, or as *spot defects*, which are random, affecting areas comparable to the single device size, and therefore more difficult to be detected (Koren and Singh 1990).

5.3.2.3 Operating Environment

Correct functioning of digital systems is based on the assumption that electrical and timing transistor parameters will remain bounded to certain margins (usually $\pm 15\ \%$). These margin tolerance specified at initial manufacturing can be violated during operating time due to shifts induced by external disturbances.

These mechanisms can produce systematic degradation overtime or abrupt failures of basic components. Transistors can be degraded due to electrical overstress and radiation whereas oxide breakdown, electrostatic discharge and ionising radiation are usually the cause of abrupt failures.

Hot carrier injection (*HCI*) has been one of the most common electrical overstress ageing mechanisms, adversely affecting both *nMOS* and *pMOS* transistors. It occurs when a charge carrier, an electron or a hole, gains enough kinetic energy to break an interface state. Different mechanisms can be responsible for *HCI* including s*ubstrate hot electrons* (*SHE*) (Ning and Yu 1974), *channel hot electrons*

(*CHE*) (Cottrell et al. 1979), *drain avalanche hot carriers* (*DAHC*) (Takeda et al. 1983) and *secondarily generated hot electrons* (*SGHE*) (Matsunaga et al. 1980).

Negative Bias temperature instability (*NBTI*) (Schroder and Babcock 2003) is also a critical reliability concern for p*MOS* transistors (not so much for n*MOS*) and has been a persistent issue for generations below *130 nm* (Schroder et al. 2003, Alam et al. 2007). Interface traps are generated during negative bias conditions $\left(V_{gs} = - V_{dd}\right)$. Higher temperatures seem worsen *NBTI*, producing larger voltage, which if maintained over long periods (*NBTI* exhibits logarithmic dependence on time), may significantly increase delays (Kumar et al. 2006, Kaczer 2005).

Another example of electrical-overstress mechanism is the *latchup* described in Sect. 4.5.1 which can also be triggered electrically (Gregory and Shafer 1973).

As described in Sects. 4.2–4.4 *non-ionising radiation* can be the cause of *DDD* while *TID* effects can be induced by ionising radiation. Other degradation mechanisms can affect interconnection logic, e.g. *electro-migration* (*EM*).

In contrast with the previous long-term degradation mechanisms, the effect of noise can produce abrupt failures. Examples of these are faults induced by the effects of noise including *oxide breakdown*, *electrostatic discharge* (*ESD*) and *ionising radiation*.

Oxide Breakdown is the destruction of an oxide layer of a semiconductor device, e.g. *time dependent dielectric breakdown* (*TDDB*), *early-life dielectric breakdown* (*ELDB*), and *EOS/ESD-induced dielectric breakdown*.

The ionising radiation mechanisms and the faults related to it have already been discussed in Sect. 4.5.

5.3.3 Phase of Creation and Occurrence of Faults

Faults that take place during the manufacturing phase are *development faults* in contrast with *operational faults* that take place during the service delivery of the operation phase (Landwehr et al. 1994), e.g. faults due to radiation as in Chap. 4.

5.3.4 Nature/Dimension

According to their nature faults can be categorised as *hardware* (such as combinational and sequential logic defects due to ageing, radiation, etc.) or *software* (branch errors, missing instructions and pointers, etc.). The scope of this work focuses exclusively on hardware faults and their effects.

5.3.5 *System Boundaries*

With respect to the system boundaries, faults can also be classified as internal (originate inside the system boundary) or external (originate outside the system boundary).

Internal faults are those that arise from within a system, often due to design flaws. These are usually repeatable for a given set of inputs in the system. In addition, they can also be the result of implementation faults, which if random, are difficult to repeat.

External faults are those that originate from outside the system, propagating into the system. These are normally the result of interference cause by the physical environment including *environmental faults* (e.g. radiation, temperature changes), accidental damage from an external system (e.g. power surges from an external power supply), etc.

5.3.6 *Phenomenological Cause*

The key components of embedded systems have an inherent susceptibility to Natural (Jennings 1990) and human-made (Hugue and Purtilo 2002) faults. Natural faults are generally random in nature and are caused by natural phenomena, without human participation. These are normally a consequence of environmental overstress. Human-made faults are the result of human action, including design and interaction faults (operational misuses), and are usually due to mistakes in the design, implementation, or use of systems.

5.3.7 *Capability/Objective/Intent*

Following the previous classification, human-made faults can either be deliberately harmful (*malicious faults*) or can be triggered without purpose and awareness (*non-malicious faults*) (Brocklehurst et al. 1994). *Accidental faults* are due to mistakes and bad decisions as long as they are not made deliberately; these include interaction, design and implementation faults. It is obvious that all natural faults have no intention and therefore are accidental. *Incompetence faults* are faults due to mistakes or bad decisions that were the result of the lack of professional competence.

5.4 Classification of Faults by Manifestation

Apart from their origin, faults can be classified based on attributes related to their manifestation, including their response, dimension, reproducibility, extent, persistence, value, detectability, etc... Table 5.3 shows a number of faults classified by their manifestation attributes.

Table 5.3 Classification of faults by manifestation

Response-timeliness (Qian 2008)	Omission faults		
	Commission faults		
Dimension	Hardware		
	Software		
	System		
Activation reproducibility (Avizienis et al. 2004)	Solid		
	Elusive	E.g. pattern sensitive faults (effects of temperature, delay in timing due to parasitic capacitance)	
Extent	Local		
	Global		
Persistence/duration	Permanent	Easiest to diagnose; once the component fails it will never work correctly again	
	Temporary	Transient	
		Intermittent	
Value	Determinate		
	Indeterminate		
Plurality	Single		
	Multiple		
Correlation	Independent		
	Related		
Damage (see Table 4.2)	Soft	Transient-soft	
		Static-soft	
	Hard		
	Pseudo-Hard		
Status	Dormant		
	Active		
Prospect (Laprie 2008)	Foreseen		
	Unforeseen		
	Foreseeable		
Seriousness	Benign		
	Malicious (Meyer and Pradhan 1991)	Symmetric	Omissive
			Transmissive
		Asymmetric (Thambidurai and Park 1988)	Transmissive (Byzantine)
			Strictly omissive (Azadmanesh and Kieckhafer 2000)

(continued)

Table 5.3 (continued)

Detectability (Pomeranz and Reddy 1993)	Detectable	Recoverable DRE (Kadayif et al. 2010; Weaver et al. 2004)	
	Undetectable	Operationally redundant	
		Unrecoverable can lead to DUE and SDC (Kadayif et al. 2010; Weaver et al. 2004)	
	Partially detectable	Under certain conditions, can be detectable and irredundant, Can lead to can lead to DUE and SDC	
Diagnosability			
Containability			
Recoverability			

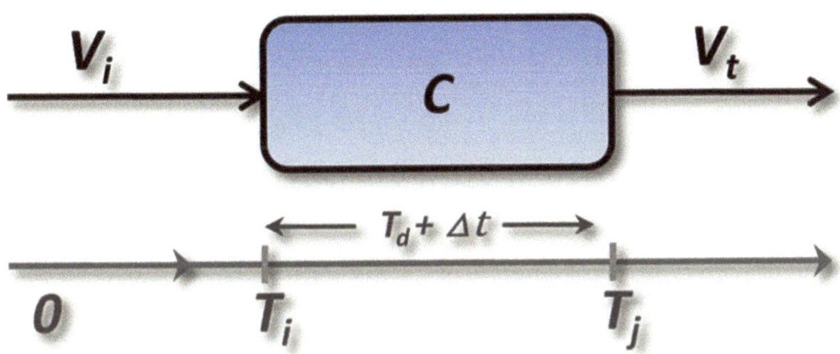

Fig. 5.3 Input-response mechanism of component C with single output

5.4.1 *Response-Timeliness*

Let a component C (see Fig. 5.3) receive a nonempty input sequence ($V_i \neq$ null), consistent with the specification, at time T_i. For V_i, the response V_j at time T_j is correct iff:

- $V_j = W_j$ at time T_j, where W_j is the expected value according with the specification and
- $T_j = T_i + T_d + \Delta t$, where T_d is the minimum delay time of the component, Δt is unpredictable delay time such that $0 \leq \Delta t \leq T_{max}$ given that T_{max} is the maximum unpredictable delay of C.

Using the previous definition of *correct response* by (Qian 2008), there can be four ways with regards to timeliness and expected value, by which a response can deviate from the specification, which leads to the following classification of faults: omission, timing, timely and commission faults:

Omission faults involve the absence of actions when these should be performed. A fault that causes a component C not to respond to a nonempty input sequence ($V_i \neq null$) is an *omission fault*. The potential resulting failure would be an *omission failure*, whose response would have the following properties:

- $V_j = \text{null}, \quad T_j = T_i + T_d + \Delta t$ and
- $V_j = W_j, \quad T_j = \infty$

A timing fault is a fault that causes a component C to respond with the expected value Wj to a nonempty input sequence ($V_i \neq$ null) either too early or too late. The corresponding failure would be a timing failure. Using the previous mathematical notation:

$$V_j = W_j, \quad \text{either} \quad T_j < T_i + T_d \quad \text{or} \quad T_j > T_i + T_d + T_{max}$$

A *timely fault* is a fault that causes a component C to respond to a nonempty input sequence ($V_i \neq null$), within the specified time interval, but with a wrong value. The corresponding failure would be a *timely failure*:

$$V_j \neq W_j, \qquad T_j = T_i + T_d + \Delta t$$

Therefore, an omission fault is also timely fault with a null value produced on time.

A commission fault of a component C is any violation from its specified behaviour, with the following properties:

$$\begin{aligned} V_j \neq W_j, &\quad T_j = T_i + T_d + \Delta t \text{ or} \\ V_j \neq W_j, &\quad T_j \neq T_i + T_d + \Delta t \text{ or} \\ V_j = W_j, &\quad T_j \neq T_i + T_d + \Delta t \end{aligned}$$

Consequently, a commission fault is a subset of all other three types of faults.

5.4.2 Consistency

Before classifying faults with respect to consistency, the definition of correct response was extended by (Qian 2008). Qian's definition and fault classification are suitable for systems that are required to produce replicated responses for a given input sequence. Examples of such systems are:

- TMR (Johnson 1989; von Neumann 1956) systems (see Sect. 3.4.1.1) a non-faulty component is required to send its output to three other components or
- A non-faulty component, which is part of some Byzantine agreement protocol, is required to send its output to other components.

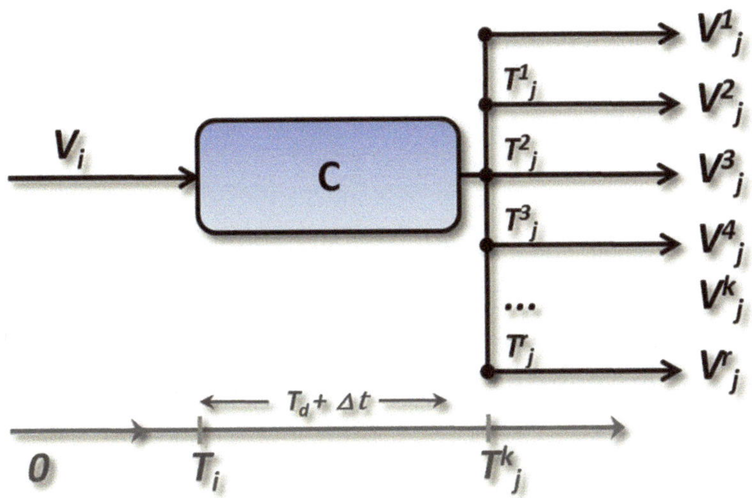

Fig. 5.4 Input-response mechanism of component C with replicated output

Figure 5.4 shows the response mechanism of a non-faulty component C with multiple r identical outputs, as a result of receiving an input sequence V_i at time T_i.

The resulting outputs are defined as:

$$V_j = \left\{V_j^1,\ V_j^2,\ V_j^3,\ \ldots,\ V_j^r\right\}$$

where $V_j^k, 1 \leq k \leq r$ are r outputs.

$$T_j = \left\{T_j^1,\ T_j^2,\ T_j^3,\ \ldots,\ T_j^r\right\}$$

where $V_j^k, 1 \leq k \leq r$ is produced at time T_j^k.

For a component C, with an input sequence V_i, received at time T_i, its replicated response is correct (*correct replicated response*) iff:

$$V_j = W_j,$$

where W_j is the expected vector of replicated outputs; and

$$T_j^k = T_j = T_i + T_d + \Delta t \text{ for all } k, 1 \ \leq k \leq r$$

Therefore, in a correct replicated response all individual responses must have the correct values "on time" (according to the specification), but not necessarily "*at the same time*". For instance, responses with same values, which take place at different times, nevertheless within the specified interval, would be part of a correct replicated response. Such interval is the *skew interval* and it is the period within which all individual responses are produced. It is defined as $\left|T_j^1 - T_j^k\right|$. For any two outputs:

$$\left|T_j^1 - T_j^k\right| \leq T_{max} \quad \text{for all } 1, k, \quad 1 \leq 1, k \leq r.$$

For replicated-response systems, two types of faults can be considered. A *consistent fault* takes place when individual responses of a component deviate from the specification in an identical manner whereas *inconsistent faults* are the ones that cause any other breach of the specification.

An incorrect replicated response is a *consistent fault* iff:

$$V_j^1 = V_j^k, \text{ and } \left|T_j^1 - T_j^k\right| \leq T_{max} \text{ for all } 1, k, \quad 1 \leq 1, k \leq r.$$

Note that a consistent fault causes a component to produce identical values (not necessarily correct values) within the skew interval, although "not on time". A few examples of consistent faults are faults with the following properties are:

- Some outputs being on time and the rest are produced early with correct values
- Some outputs being late and having correct values but the rest are correct
- All outputs having identically incorrect values.

An inconsistent fault is an incorrect replicated response iff its individual responses do not satisfy the consistent failure conditions explained above. A *byzantine fault* (the behaviour of a component that gives conflicting values to other components) is a type of inconsistent fault.

5.4.3 Maintainability: Detectability, Diagnosability and Recoverability

Fault detection, diagnosis and recovery are required to ensure resilience. Testing and diagnosis may be online, offline or a combination of both (Kaegi-Trachsel et al. 2009). Online testing and diagnosis are performed concurrently with system operation whilst offline methods require that the system or subsystem is taken out of service for a specific time.

Often, online testing is used for detection while offline diagnosis locates and identifies the fault(s). As soon as the system/subsystem is repaired, offline testing can be used to verify that the repair was successful before placing it back to normal operation.

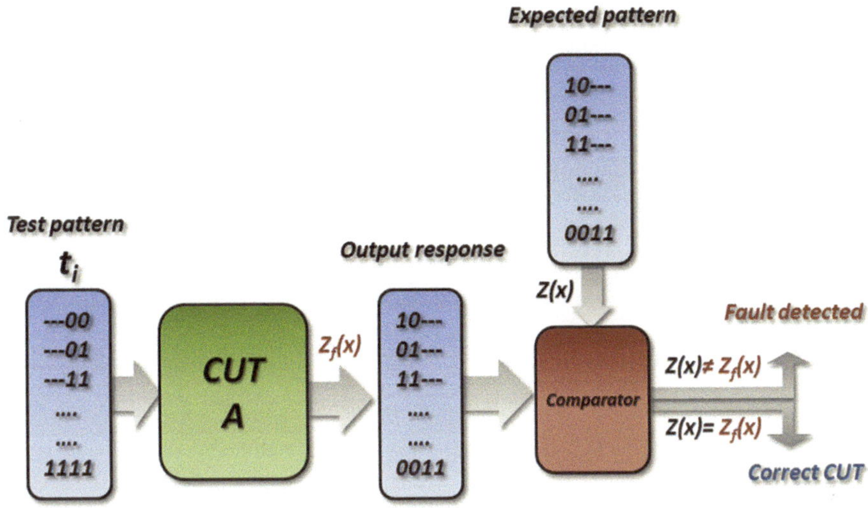

Fig. 5.5 Basic testing flow of circuit under test (CUT)

Test vectors are used by automatic test pattern generation tools (ATPG) (Agrawal and Chakradhar 1995; Roy et al. 1988) to attempt the detection of all or most modelled fault groups. A *test vector* is a string of n logical values (0,1, or irrelevant X) that are applied to the N corresponding primary inputs (PI) of a circuit, at the same time frame, in order to detect one or more faults (Roy et al. 1988).

The specification of a test vector should have two components: the input to be applied and the expected fault-free output (e.g. $t = I/O = 0010/11$). A fault will be detected if the output under fault is different than the expected output. If a series of test vectors are applied in a specific order, the term *test sequence* is used, otherwise it is a *test set* (Roy et al. 1988). The term *test pattern* is often used to refer to any of these three terms.

The testing process involves test pattern *generation* (ATPG), test pattern application in the CUT and output evaluation by an output response analyser (ORA) (Stroud 2002). Figure 5.5 shows the basic testing flow of a circuit, whose output response after processing a test pattern is compared to the expected pattern (fault-free response pattern that a non-faulty circuit would exhibit). The quality of testing will depend on its fault coverage (defined in Sect. 2.5.2.3) and speed.

Following (Abramovici et al. 1994), let x be a random input vector, and Z(x) the function of a circuit under test A with an input x. A fault f would transform A into a new circuit Af with function Zf(x). Let T be a test set $T = \{t_1\ ,\ t_2,\ t_3 \ldots\ t_n\}$ formed by n ti test vectors where ti $\geq$ 1. Figure 5.5, the CUT A is tested by applying T and comparing the output response Zf(ti) with the expected pattern Z(ti). A fault is detected if the output response is different than the expected pattern:

$$Z(x) \neq Z_f(x).$$

With regards to fault diagnosis, a test is said to distinguish two faults f_1 and f_2 (*distinguishable/diagnosable faults*) if the output response of the faults are different from each other:

$$Z_{f1}(t) \neq Z_{f2}(t).$$

Conversely, two faults are functionally *equivalent* if all tests that detect f1 also detect f2:

$$f_1 \sim f_2 \quad \text{iff} \quad Z_{f_1}(t) = Z_{f_2}(t) \quad \text{for all} \quad t.$$

Functional equivalence can be easily analysed for logic gates. An example of equivalent faults is shown in Fig. 5.6a. There is not an existing test that can distinguish between the *s-a-0* faults occurring in the input and output of the *AND*. The same applies for all s-a-1 faults that occur in an *OR* gate.

Figure 5.6b shows six faults, two of each class A, B and C, in a CUT. Note that detection, diagnosis together with containment and recovery are some of the goals of testing (as specified in Sect. 2.5.2.3).

For fault detection at least one vector is needed (fault detection provides only whether the circuit is free of faults or not).

For fault diagnosis at least one vector that produces different responses for every fault class (fault diagnosis aims to determine time, location and type of the detected fault) is needed.

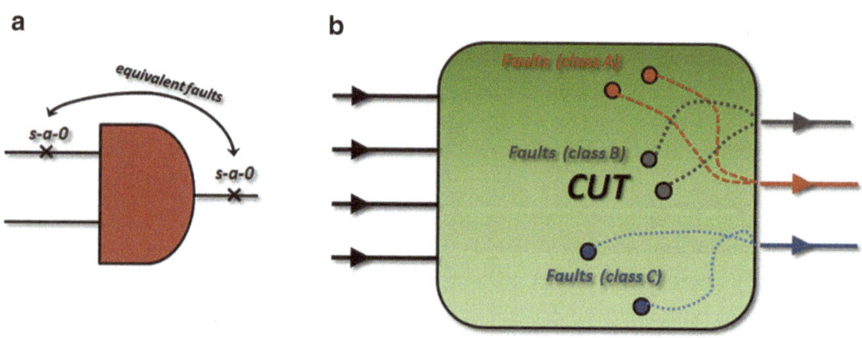

Fig. 5.6 Fault diagnosis and equivalent faults. (**a**) Example of equivalent fault. (**b**) Fault detection and diagnosis vs. vectors

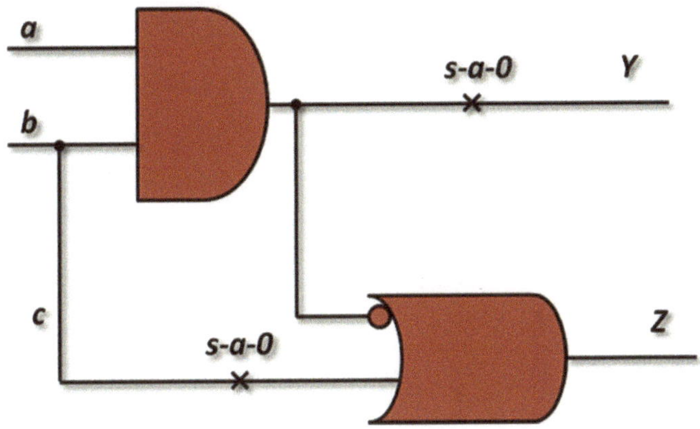

Fig. 5.7 Example of non-diagnostic detection equivalence

For combinational circuits with multiple outputs, two types of equivalence can be differentiated (Sandireddy and Agrawal 2005):

- Diagnostic equivalence: two faults are f_1 and f_2 are diagnostically equivalent iff the functions of the two faulty circuits are identical at each output
- Detection equivalence: two faults are f_1 and f_2 are detection equivalent iff all tests that detect f_1 also detect f_2, not necessarily with the same output.

Figure 5.7 shows an example circuit with two single *s-a-0* faults in the input *c* and output *Y* lines. Both are detection equivalent faults but are not diagnostically equivalent.

A fault f_2 dominates f_1 ($f_2 > f_1$) if the test set for f_1 (T_1) is a subset of the test for f_2 (T_2). All tests pattern of f_1 would detect f_2. Therefore, f_1 implies f_2 and including f_1 in the fault list would be sufficient.

If two faults dominate each other then they are equivalent:

$$f_1 \sim f_2 \quad \text{iff} \quad f_1 > f_2 \quad \text{and} \quad f_2 > f_1.$$

ATPG tools generate test patterns that target possible physical faults according to the fault model (Agrawal and Chakradhar 1995; Roy et al. 1988). An increase in the complexity of circuits involves bigger fault dictionaries and patterns, slowing down the ATPG process.

The implementation of quick detection and diagnostic mechanisms can improve the effectiveness of resilience. One way of creating compact sets a is fault collapsing, which is the process of reducing the number of faults by using redundancy, equivalence and dominance relationships among faults is called fault collapsing (Abramovici and Breuer 1979). To lessen the burden of test generation, two main types of fault collapsing are used:

a

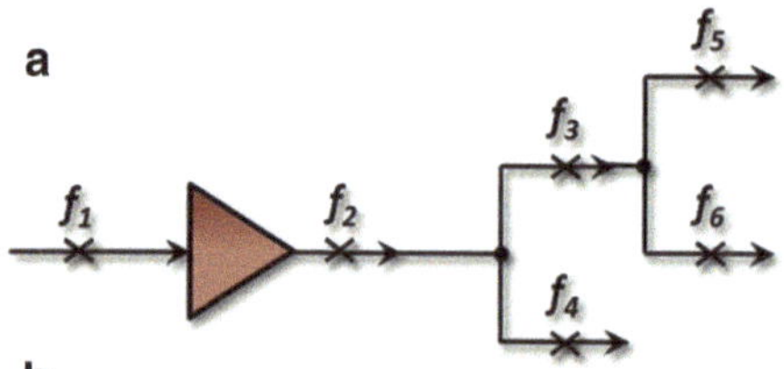

b

Line 1	Line 2	Equivalence	Dominance
f_1	f_2	Y	Either
f_2	f_3	N	f_2 over f_3
f_2	f_4	N	f_2 over f_4
f_3	f_5	N	f_3 over f_5
f_3	f_6	N	f_3 over f_6
f_5	f_6	N	N
f_2	f_4	N	f_2 over f_4
f_3	f_4	N	N

c

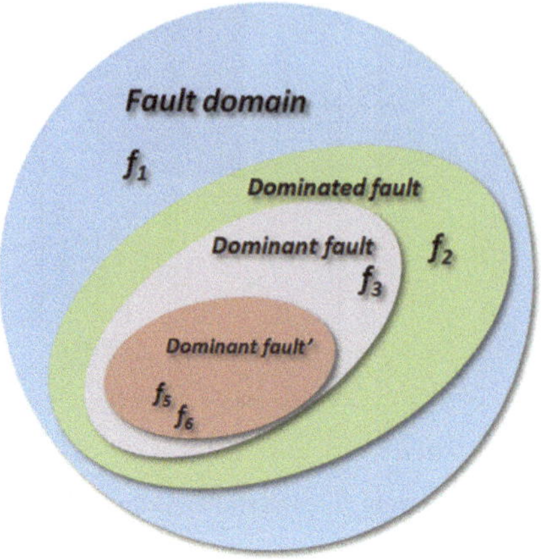

Fig. 5.8 Example of non-diagnostic detection equivalence

- *Fault equivalence collapsing*: uses the notion of fault equivalence to remove most of the equivalent faults from the pattern. Faults of a logic circuit can be divided into *N* disjoint equivalence subsets S_i, where all faults within a subset are mutually equivalent. A fault set S_i is *collapsed* if it contains one fault from each equivalence subset.
- *Fault dominance collapsing*: uses the notion of fault dominance to remove dominating faults from the equivalent collapsed faults. If fault f_2 dominates f_1, then f_2 is removed from the fault list.

Figure 5.8b shows the dominance and equivalence relationships of a given circuit (Fig. 5.8a).

Both equivalence and dominance relationships are transitive. For instance, Fig. 5.8c—if fault f_2 dominates f_3 and f_3 dominates f_5 then f_2 will dominate f_5. Collapsing algorithms use this transitivity property to reduce the fault patterns (Prasad et al. 2002). If fault detection is the only objective (e.g. fail-safe system that do not require diagnosis), then fault dominance collapsing can be used to further reduce the fault list.

A fault *f* is *detectable/testable* if there is a test *T* that is able to test/detect *f*, otherwise *f* is an *undetectable/untestable* fault (Agrawal and Chakradhar 1995). Undetectable faults can be partitioned into two subsets: partially detectable faults and redundant faults. A test set/sequence is said to be *N-detectable* if all faults are detected at least N times with N different test vectors (McCluskey and Tseng 2000). The higher the value of N the higher is the fault coverage.

A circuit is *redundant* if the function realised by the circuit without fault(s) is the same as the function realised by the circuit with one or more faults (Carter 1979). *Redundant faults* are undetectable faults that do not affect the circuit operation (*operationally redundant*). It can be argued why is it of interest to discover redundant faults if they do not affect circuit logical behaviour. Discovering and removing redundant faults from the tests is important for the following reasons:

- In redundant circuits, the presence of undetectable faults can invalidate certain tests, raising problems such as (Friedman 1967):
  - If f_1 is a detectable fault and f_2 is an undetectable fault, then f can become undetectable in the presence of g. In that case, f_1 is a second-generation redundant fault (Friedman 1967)
  - If two undetectable single faults g_1 and g_2 are simultaneously present in a system, then they can become detectable
- Due to time and power consumption constrains, it is not feasible to perform a complete search of all possible faults in any given circuit. There are certain needs for minimising the tests patterns that detect existing faults.
- Additional current drains can be induced due to redundant faults such leakage faults (Mao et al. 1990; Xiaoqing et al. 1996) and gate oxide shorts (Hawkins et al. 2003; Segura and Hawkins 2005), which are especially undesirable in low-power devices.
- Redundant defects may indicate a latent reliability problem.

In the case of combinational circuits, all undetectable faults are redundant faults (Abramovici and Breuer 1979). Testing sequential logic is significantly more difficult that testing combinational logic, whose response is a function of its initial state.

In the case of sequential circuits, it has been shown that certain faults for which a test sequence does not exist, under certain conditions, faulty behaviour may be detected. These are *partially/potentially detectable faults*, faults that affect circuit operation under some states (*partially irredundant*), but are not manifested at the outputs for any input sequence under other states (therefore undetectable) (Pomeranz and Reddy 1993).

The testability features, observability and controllability (defined in Sect. 2.5.2.2) are important to increase the effectiveness of fault detection. Therefore, observability and controllability would increase the number of testable and tested faults and so the fault coverage. Conversely, lack of testability would increase the number of untestable and untested faults.

In terms of diagnosis, different approaches can be followed: *offline diagnosis* based on fault dictionaries, effect-cause diagnosis also called *online* or *dynamic diagnosis* based approaches, or a combination of these two approaches (Smith 1997).

Finally, as in Table 5.3, faults can also be classified according to their diagnosability, containability and recoverability.

5.5 FT and System Modelling

When a fault appears, extra redundancy is needed to cope with it. Thus, redundancy and ability to use it form a sort of combination that is required to be applied within kind of framework for implementation of reconfigurability.

When system and fault modelling are developed together, system behaviour in presence of faults and control process of FT can be considered at earlier stages, taking into account mutual dependencies of solutions at every stage of a process of design and development.

Regarding embedded systems we are facing constrains and very often mutually exclusive requirement in reliability, performance and power consumption. This situation limits design options as Fig. 5.9 illustrates.

At the same time, redundancy can be used for various purposes and can be involved as essential part of reconfiguration of the connected computers. System reconfiguration purposes are:

- Performance improvement
- Reliability enforcement
- Energy-wise use

The approach of developing reconfigurable computer was named PRE-smart computers (Schagaev 2009; Schagaev et al. 2010, 2013). Possible inheritance of properties is shown on Fig. 5.10.

Thus, PRE (performance-, reliability- and energy-) wise systems might be designed rigorously, using reconfigurability and recoverability as system features introduced at conceptual level. Success of PRE designs use of this approach for connected computers (networking, clusters multiprocessing) depends on careful trading of the redundancy introduced to achieve the required property.

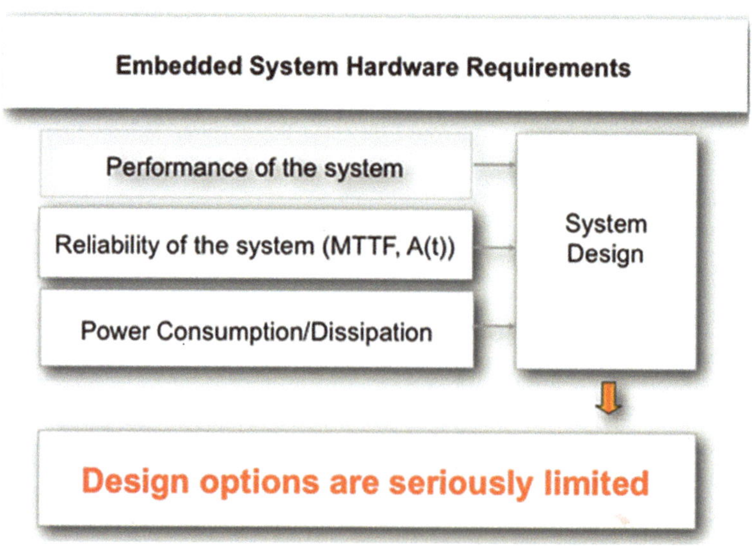

Fig. 5.9 Performance, reliability and power concerns on the design of embedded systems

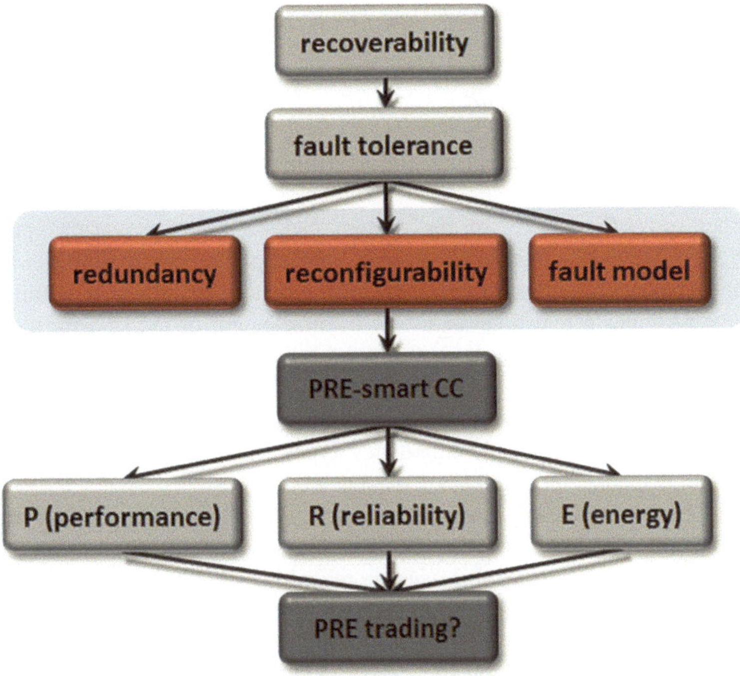

Fig. 5.10 Reconfiguration purposes of a system including fault tolerance

5.5.1 *Trading P, R, E*

Describing a system in terms of Structure, Information and Time as parameters of redundancy (Schagaev 2001) it is possible to estimate reliability, as well as other properties. Redundancy might be weighted, say, in units or values with or without relation to the steps of any supportive algorithm that applies it to form and control configuration.

While time and information is clear in units—seconds and bits, the structure, especially structural redundancy requires some extra efforts. Note also that time, information and structure are considered as independent variables. Structural redundancy for our purposes might be measured using graph or graph notation:

$$dS :< dV, dE >$$

where *dS* denotes introduced structural redundancy, while dV and dE denoted extra vertices and edges added in the structure to make step of algorithm implemented.

Then our efforts toward PRE goal can be measured quantitatively as redundancy vector:

$$dR = < dT, dS, dI > .$$

Time, information and structure as mentioned above are considered as independent variables.

Further, reconfigurability of the system can be used for various purposes (Fig. 5.9). To be able to use redundancy and apply reconfigurability we need consider introduction of supportive tool for reconfigurability a syndrome. The syndrome provides a real-time status for every element of the system, it show a "snapshot-status" of the system in terms of reliability, performance and power awareness.

Let us now analyse how generalised algorithm of fault tolerance might be implemented using mentioned redundancy types and reconfiguration introduced as a system property.

As Fig. 5.9 shows almost mutually contradictive requirements of performance, power consumption and reliability limit design choice in design of embedded systems. These constrains become even more serious when system is implemented on a chip (SoC). In this case reliability of elements, performance and power consumption are defined and limited by the same technology.

Regarding reliability available redundancy hints a potential solution: it is known that max (P_i,P_j,P_k,P_l) is achieved when all mentioned probabilities of independent elements are equal.

For example, for SoC it technologically means that we need to introduce in the system different redundancy levels for each mentioned group of elements to equalise reliabilities of various group. The same statement is true for performance and power consumption, the latter is required to avoid "periferial" overheating of elements from neighbouring subsystems. This segment of research needs attention and is worth further exploring.

5.5.2 *GAFT: Generalised Algorithm of Fault Tolerance*

It is well known that fault tolerance of a computer system can be achieved by introduction of static redundancy in hardware and system software (HW/SSW). It is also well known that using traditional approaches of fault tolerance (Anderson and Lee 1981, p. 81; Avizienis 1971; DeAngelis and Lauro 1976) is expensive in terms of time, information or hardware overheads.

To avoid this, the authors (Schagaev 1986a, b, 1987; Sogomonian and Schagaev 1988) proposed to consider a fault tolerance not only as a feature but also as a process that can be implemented algorithmically.

The introduction of static redundancy in HW and SSW might be prohibitively expensive in terms of cost and power consumption. Therefore, a process that

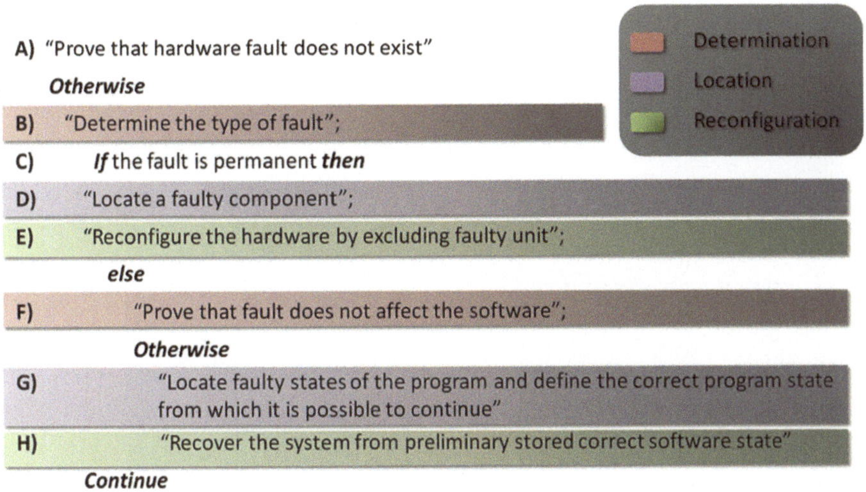

Fig. 5.11 GAFT: generalised algorithm of fault tolerance

implements fault tolerance assuming dynamic interaction of existing redundancy types between elements (Fig. 5.11) can tackle these problems. The three-step algorithm (Sogomonian and Schagaev 1988) has been further developed. The outcome of this work is the Generalised Algorithm of Fault Tolerance (GAFT), a five-step fault-handling algorithm (Fig. 5.11):

- Detecting faults
- Identifying faults
- Identifying faulty components
- Hardware reconfiguration to achieve a repairable state
- Recovery of a correct state(s) for the system and user SW.

The different types of redundancy (information, time and structural either hardware- or software-based) can be used to implement every step of GAFT.

At the same time, there is an essential extension for this algorithm regarding the role of system software in implementation of hardware fault tolerance. Bullet (F) of GAFT Fig. 5.11 above has a crucial importance, especially due to latency of hardware faults that is increasing with complexity of computer systems. System software related part of GAFT should be extended as Fig. 5.12 proposes. System software should find correct state of software first to be able to continue recovery.

Only and when hardware recovery and, if necessary, reconfiguration have been done the systems software fault-free state should be found and procedure of toleration of fault completed. When latency is zero the latest recovery point is sufficient. In real world we do not know what is size of latency is; therefore, a searching of correct state of software is unavoidable. Fortunately the problem of finding of software correct state was solved (Schagaev 1986a, b, 1987).

Generally speaking, the more complex the system implementation is, the more complex Fault detection and diagnosis will be. This is particularly true for

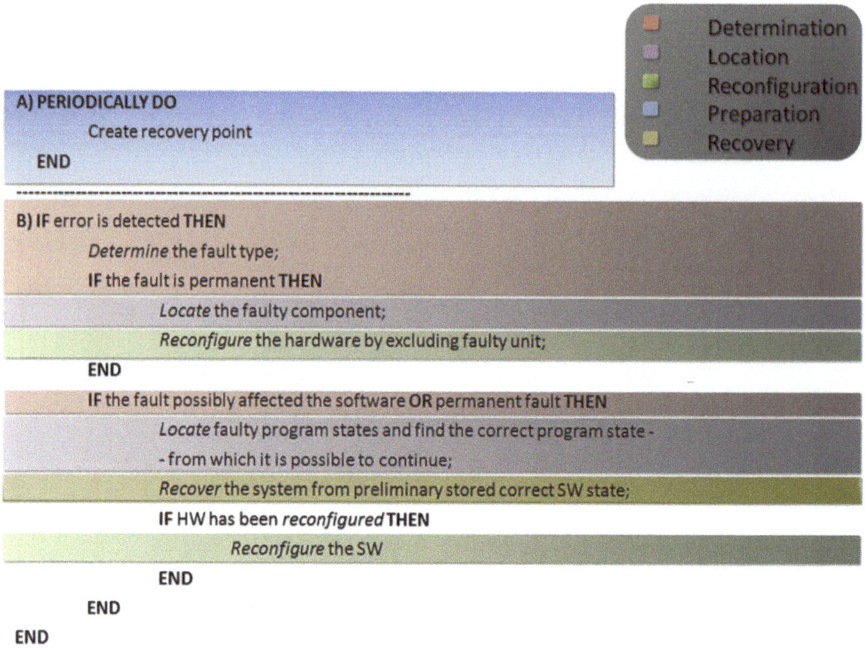

Fig. 5.12 GAFT: generalised algorithm of fault tolerance

multicore system (MIMD) processors that support vector instructions (SIMD) and pipelined implementations, which are particularly complex. In principle the complexity of GAFT implementations also depends on the complexity of the system, its faults and fault tolerance models.

Hardware support is a faster mechanism than software support to achieve reliability and with a proper design there should be little performance degradation. However, the introduction of static redundancy in HW might be prohibitively expensive in terms of cost and power consumption. Therefore, a process that implements fault tolerance assuming dynamic interaction of existing redundancy types between elements can tackle these problems.

GAFT might be used for comparison and overview of different design solutions of FT systems. It also allows controlling fault coverage at every step of system design, providing a tool for the selection of efficient solution and estimation of overheads.

A substantial redundancy can cover multiple steps in GAFT, such as fault detection and fault recovery; take, for example, TMR systems, when a faulty channel output is overruled by two correct outputs.

The implementation of the hardware checking (step A of GAFT) at different levels causes different timing for GAFT completion: microseconds for the instruction level, milliseconds for the procedure level, hundreds of milliseconds for the

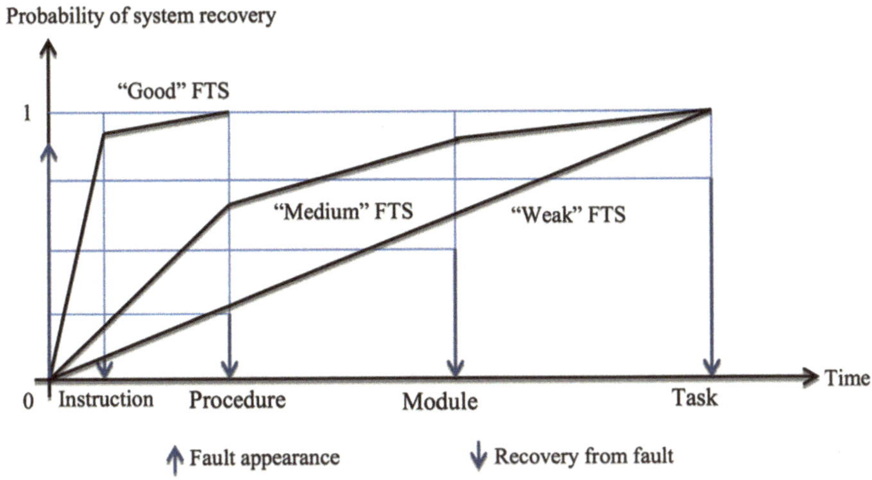

Fig. 5.13 System recovery time according the level of implementation of checking and recovery schemes

module level, seconds to tens of seconds at the task level and tens of seconds to minutes or even hours at the system level.

Different implementations would have different properties in terms of timing, fault coverage, types of faults that can be tolerated, power consumption, complexity and cost. Figure 5.13 illustrates various solutions of fault toleration at the level of instruction, procedure, module and task. All shown systems are fault tolerant, have different cost, redundancy level involved and coverage of faults tolerated.

It is, therefore, wise to combine different checking and recovery schemes in one system to achieve required specifications, it is wise to combine different checking and recovery schemes in one system.

For example, it might be beneficial to protect the processor and the memory by using hardware schemes at the level of instructions (duplicated processors, triplicate memory) and use higher-level schemes (procedure or module) for the other hardware components due to cost and power constraints.

Processor and memory is used at every instruction execution. The implementation levels are not mutually exclusive; for example the combination of hardware- and software-based checking can significantly improve fault coverage. In general, the higher the implementation level the less hardware support is required, but with higher timing and software coding overhead.

A good fault tolerant system tolerates the vast majority of transient faults within the interval of instruction execution, making them invisible for other instructions (and software). At the same time, our assumptions about transient "live" fault is arbitrary, and thus, transient faults with longer time range or permanent faults might be detected and recovered differently, for example, at the procedural or task level of system software.

Taking into account that transient faults occur at an order of magnitude more often than permanent faults, transient fault tolerance must be done extremely effective.

In turn, it is necessary to implement special schemes for HW reconfigurability and recoverability to eliminate the impact of permanent faults on the system.

GAFT completion requires three fundamental processes, called in literature P1, P2 and P3 [Stepanyants 01]. The error *checking process* P1 is responsible for checking the system state in terms of hardware fault existence/appearance. The second process, the *error recovery process* P2 prepares recovery states when it is scheduled by the system. The third one, the *functional process* P3 is a process of calculations or instruction executions.

When a transient fault is detected, its toleration assumes recovering the information that has been modified and restoring hardware states. The third one, the *functional process* P3 is the process of calculation or instruction execution. Let us assume the primary function of the real-time critical system as "process three" or P3.

Therefore, if the system ensures full functionality and transparent application recovery for the process P3 (from a predefined set of faults in a given time frame) then the system is fault tolerant.

In other words, *we define a system as fault tolerant if and only if it implements GAFT transparently for applications.*

Or, in a bit more details:

If the system ensures full functionality and transparent application recovery for the process P3 (from a predefined set of faults in a given time frame) then the system is fault tolerant.

That is, *we define a system as fault tolerant if and only if it implements GAFT transparently for applications.*

5.5.3 GAFT: System Estates and Actions to Implement Fault Tolerance

GAFT above presents an approach of achieving a new and complex property (fault tolerance) considering it as a process with several steps and phases. Further generalisation and detailed analysis of system state change is presented below.

At any given time, a single processing element (SPE) system can be in one and one only of five possible states: ideal, faulty, erroneous, degraded and failed. Figure 5.14 shows the five S states, the potential T transitions and the M mechanisms involved in fault tolerance. Two different areas can be differentiated: the green area at the bottom half represents the conventional environment with no FT capabilities whereas the red area at the top half represents the possible states and transitions in a dependable environment.

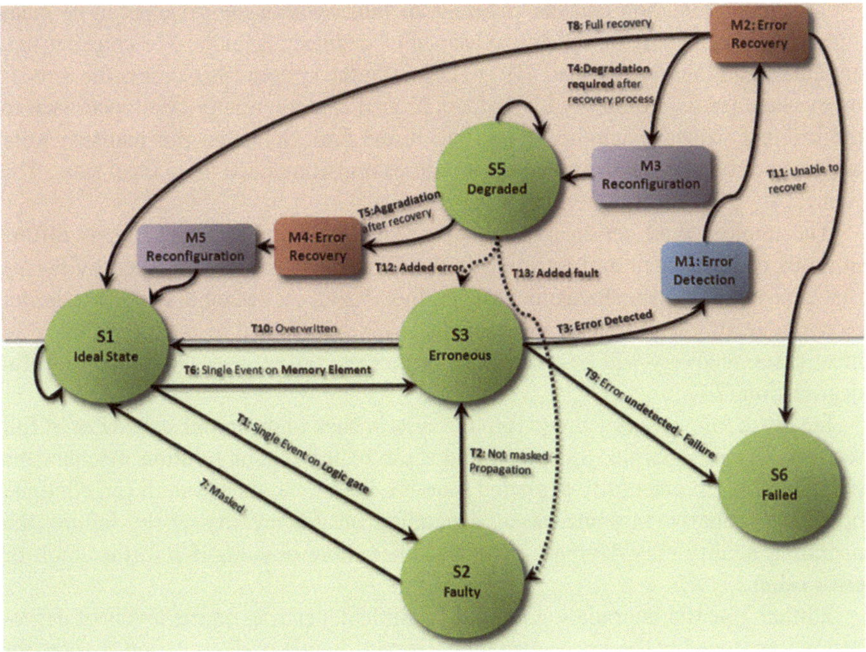

Fig. 5.14 GAFT: generalised algorithm of fault tolerance

Considering S1 as the initial state, a single event can change this ideal state. Note that with the exception of the transition T12, Fig. 5.14 does not contemplate the cases where a single event can occur between transitions. If the deviation in the form of voltage transient introduced by such event affects combinational logic, the system would turn (Transition 1) into a faulty state S2. The voltage transient may propagate to sequential logic such a memory cell or latch, potentially flipping the bit and contaminating the data flowing within the system, and leading the system to an erroneous state S3 (soft error in Fig. 5.14).

However, there are three masking effects that can prevent transition for this particular event from S2 to S3: logical masking, electrical masking and latch-window masking. Logical masking happens when one of the other inputs of the affected gate is in controlling state so that the output does not vary (e.g. 1 for an AND gate, or 0 for a NAND gate).

Electrical masking happens when the voltage transient impacts successive logic gates and propagates through the logic chain fading out before reaching the registered output. Latch-window masking happens when the arrival of the pulse is outside the latching window, usually based on the setup time and hold time of the sequential logic.

Nevertheless, if during the ideal state S1, as a result of the single event the voltage transient affects the sequential logic directly, the system state would transit straightforward to an erroneous state S3 (transition 6).

The implementation and the coverage of faults within the system can be measured probabilistically, assuming existence of undetected faults. We consider that an undetected fault would lead (T9) to failure in the system (S6), unless the error is overwritten [(e.g.: a memory bit that has flipped can potentially be flipped back to the original value by another event before the fault detection mechanisms were activated or before the error leads to a failure) transiting back to an ideal state. The probability of overwritten errors is very small].

The implemented action of fault detection and recovery mechanisms differs in terms of permanent and transient faults. Faults are initially detected by the by the fault detection mechanisms (M1 in Fig. 5.14). A detected fault that is not recoverable by the recovery mechanisms (M2) would lead to a failure (T11). In most cases, recovery will be possible in two forms: full recovery (T8) and graceful degradation (*T4*).

Ideally, a full recovery would turn the system back to the initial state *S1* or, if full recovery is not possible, SSW will make use of the reconfiguration mechanisms (*M3*) to turn into gracefully degraded state S5. In this state the system can continue operating properly. In some cases, depending on the severity of the failure, the operating quality may decrease. This becomes more obvious if a further fault or error occur.

Further graceful degradation may be possible depending on the levels of degradation introduced in the implementation. Good example of degradation support for memory can be found in (Bernstein et al. 1992, 1993).

Recovery from a degraded state takes place once the deviation has been corrected. The recovery mechanisms should be able to return the system back to correct state using additional reconfiguration (M5).

Clear, some logic framework, holistic principles to follow might help to design and develop efficient architecture with required properties. Briefly the next section explains the principles chosen for this work, as it is introduced in (Schagaev et al. 2010); these principles are subject of next chapter.

5.6 Conclusion

- The chapter introduces a detailed fault model description and analyses it together with generalised methods of fault toleration.
- We extend the model of faults defined by (Avizienis et al. 2004) and provide a classification suggesting methods for recognition and reaction.
- Manifestation, detectability, diagnosability and recoverability are discussed as one consistent flow proposing adequate solutions for diagnosis and recovery.
- We introduce the principle of reconfiguration of the system and consider how reconfiguration might be used for various purposes addressing requirements and implementing monitoring processes of:

  - Performance
  - Reliability and
  - Energy wise gain
- A generalised algorithm of fault tolerance is developed further with a full explanation of system state changes and the actions required to be implemented.

Chapter 6
Hardware Support of Resilience

6.1 ERA Concept, System Design and Hardware Elements

The development of this new architecture follows the holistic principles proposed by the ERA paradigm (Schagaev et al. 2010): simplicity, redundancy, reconfigurability, scalability, reliability and resource awareness.

To support those principles a new hardware architecture (HW) and system software (SSW) have been developed. A brief introduction to these principles is as follows:

Simplicity: Complexity is difficult to implement and handle efficiently. In addition, big complex systems are more prone to faults, thus lowering reliability.

Reliability: The highest reliability of individual components is preferable but always keeping in mind the cost-efficiency of its implementation.

Redundancy: Deliberate introduction of hardware and software redundancy provides the required level of reconfigurability to reach performance and reliability goals.

Reconfigurability: Apart from the simplicity, reliability and deliberate introduction of redundancy, it is essential to achieve balance between performance, reliability and power. Reconfigurability serves three main purposes: performance, reliability and power awareness. It allows the system to adapt twofold: first to recover from a permanent fault and second adjusting the requirements of the running application.

Scalability: Scalability should be kept in mind when designing a system so that it can be extended if the requirements change.

Power-awareness: Mission critical systems have significant limitations of hardware resources and power consumptions constraints (e.g. battery life). Thus, for wise resource use, reconfigurability must be introduced.

By following these principles the processes of design and development of new architecture can be defined.

V. Castano, I. Schagaev, *Resilient Computer System Design*,
DOI 10.1007/978-3-319-15069-7_6

Figure 6.1 shows a computer system as three semantically different (from the point of view of information processing/transformation) elements. The above-mentioned principles might be applied at the level of each element enabling design by far more efficient computer system.

In terms of information processing point of view the hardware is based on a Single Processing Element (*SPE*) that is divided into three areas (Fig. 6.1): first, the information transformation area—further called active zone (*AZ*) and second the information storage area—called passive zone (*PZ*). The interconnection of these zones is the interfacing zone (*IZ*).

All three zones must have different properties and therefore need different redundancy mechanisms to tolerate faults and make system reconfigurability possible and efficient. The proposed structure of each zone is shown in Fig. 6.2.

Active Zone: The active zone consists of the arithmetic unit and logic unit, both separated for better fault isolation and easier implementation of hardware tests.

Interfacing Zone: This includes all communication components such as the processor, the memory buses and the reconfiguration logic. A configurable bus

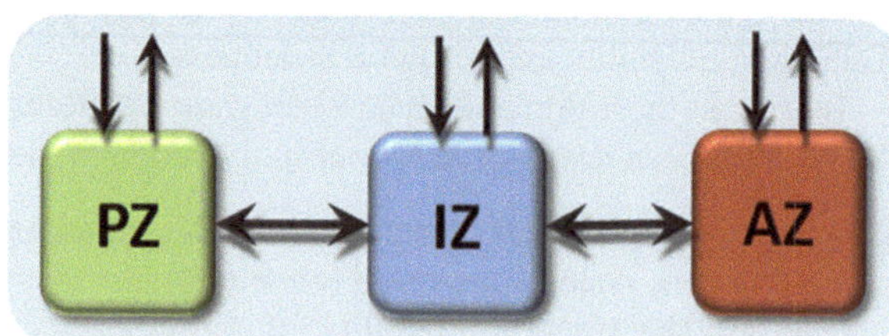

Fig. 6.1 System zones from an information processing point of view

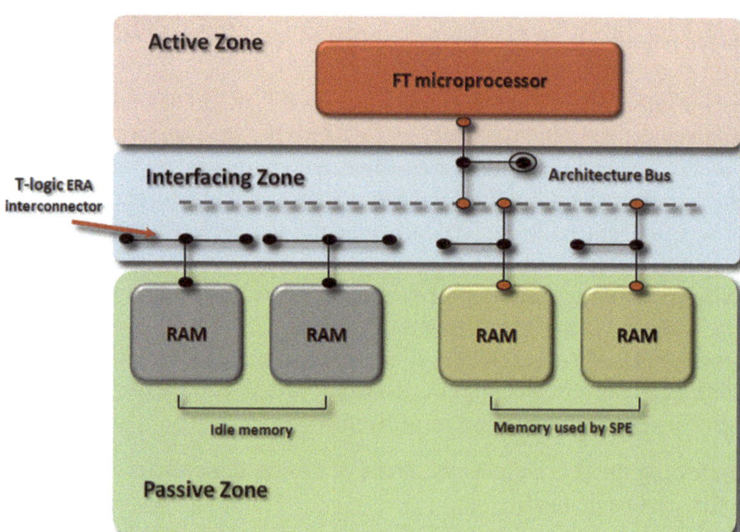

Fig. 6.2 System zones from an information processing in ERA

allows the reconfiguration of the hardware to exclude failed hardware components and go into a degraded state, or replace the failed component with a working one.

Passive Zone: This includes basic storage systems, such as memory, that do not act by themselves but are handled by controllers or devices.

Minimum deliberate redundancy has been introduced in the form of buffer, register files, replicated memory modules, majority schemes (in terms of HW) and interfacing logic. System software extra elements required to support fault tolerance are: checkpoint monitor, recovery point monitor, process synchronisation and reconfiguration monitors. They named monitors to express their uninterruptable mode of operation.

As mentioned above, hardware can be considered as three zones, (see Figs. 6.1 and 6.2) All elements in these zones have to be reconfigurable for their own purposes as well as other zones requests.

All zones might have different reconfiguration properties. Reconfiguration might have internal and external reasons. For example, when the system forms a configuration for a task execution it might deliberately and *externally* exclude some hardware elements from configuration due to task requirements or transient/permanent fault. In turn, checking schemes activates *internal* reasons for reconfiguration.

Interactions between zones define the level of reconfigurability and flexibility of the architecture. These new hardware reconfiguration abilities must be reflected and supported as new features of the architecture.

6.2 ERA Hardware Configuration: ERRIC

6.2.1 Active Zone

The main principle used in the design of the processor is simplicity. The cost of P1 and P2 implementation of GAFT depends on the structure of the processor and might become prohibitively high. Antola et al. (1986, p. 1) proved that the overheads necessary to make a CPU fully fault tolerant might easily exceed 100 %. Thus, to avoid duplication, there is a requirement to keep redundancy level needed to implement fault tolerance, as low as possible.

Following the simplicity principle, the instruction set and its implementation within the processor is reduced to the absolute minimum required to support general purpose computing. This allows the careful introduction of redundancy that implements error detection, diagnosis and recovery features. Complicated memory addressing instructions are omitted, as they are not essential.

The instruction set architecture (ISA) consists of only 16 instructions with only two of them for memory access. Such a simplified instruction set generally requires less hardware (the control unit in particular), which increases, by design, the reliability of a single processor.

Performance might also be increased as operating clock frequencies can be improved. Extra details on the instruction architecture are explained further in Sect. 7.2.

The proposed microprocessor offers clear advantages in comparison to CISC architectures: fewer and simple addressing modes, hardwired design (no microcode), fixed and simple instruction formats. In turn, a simple instruction format allows fetching of two 16-bit instructions per machine cycle. The execution steps are similar to other RISC processors.

There is no pipelining and all steps of one single instruction are executed within one memory cycle. Pipelines are one of the most vulnerable elements of modern microprocessors. The relative amount of area of the chip dedicated to pipelines is increasing with scaling design complexities. For instance, instructions can stall in the instruction queue and the longer they reside there, the higher the chances of getting struck by an energetic particle.

A transient fault in a latch or a memory cell within the pipeline (e.g. SET or SEU) can propagate and become an error at the micro-architectural level (e.g. Register file or Instruction Register). Consequently, the effects of ionising radiation in this area can lead to SEFI's, severely decreasing the overall reliability.

The absence of pipelining or caches greatly simplifies the processor design, which, in turn, simplifies the implementation of fault tolerance. A careful introduction of redundancy for checking and recovery allows the processor:

- To detect transient faults during the execution of the instruction.
- To abort the current instruction and,
- To re-execute it, all transparently to software.

There are known arguments that simple ISA's do not have enough instructions to perform most of the application operations. These argue that a small ISA would result in higher compilation effort/time, and that the resultant programs would be bigger, which would increase the amount of memory needed.

However, complex functions are well handled by the compiler instead of having specialised instructions within the ISA dedicated to very particular tasks. Regarding the size of the programs it is worth mentioning that:

- A bigger ISA such as CISC involves longer operation codes which, in turn, increase the size of programs.
- In an architecture based on a smaller ISA, register references require fewer bits.
- It is usually claimed that smaller ISA requires more memory. ERA architecture has unique design property: the size of all instructions is 16 bits, while the size of a word is 32. Thus, the instruction density per word is 2 and the previous arguments above about bigger code size for RISC do not stand to ERA.
- Since the price of the memory is very low and keeps decreasing this argument also becomes less important.

The processor has a large register file with 32 general purpose registers with a width of 32 bits and no restrictions on their use, which simplifies software development (see Fig. 6.3). All standard instructions expect exactly two arbitrary

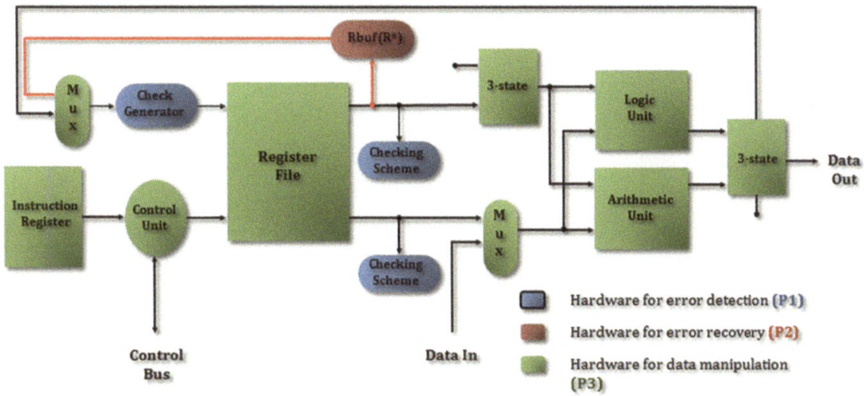

Fig. 6.3 Architecture of active zone

registers as input, and save the result of the operation in one of these two registers, thus overwriting one of the input values.

The impact of the size of the register file on overall performance of processor is a question of further research.

Memory access is only possible for 32-bit words at a time, and it has to be aligned. The main structure of the processor architecture is illustrated in Fig. 6.3.

The instructions are fetched from memory into the instruction register and decoded by the Control Unit, which also manages the execution of each instruction. Operands for each instruction being executed are fetched from either the register file or memory multiplexed into the Arithmetic or Logic Units, AU and LU, respectively.

The output data from AU or LU goes either to memory through the data bus or is written back to the Register File. The current value and type of the data might also indicate an address for branch instructions.

The hardware for fault detection and error recovery process P1 and P2 are marked in blue and red respectively. The hardware for the data manipulating process P3 is marked green.

The simplified architecture Fig. 6.3 presents allows implementation of GAFT at instruction level with reasonably small reliability (13 %) (Schagaev 2008).

As before, the three processes are essential for the guaranteed and successful execution of each instruction. Two processes P1 and P2 cope with fault checking and recovering of transient errors respectively.

P1 is initialised at the start of the instruction execution; P2 is required and initiated when fault has been detected, but it is essential that the premodified state is stored at the start of execution of every instruction.

P1 and P3 can operate concurrently. P1 and P2 have an influence on each other: the higher the fault detection coverage achieved by P1, the more successful recovery should be (Stepanyants 2001).

When data is written into the Register File, the Check Generator (marked in blue in Figs. 6.3 and 6.4) generates the checking information for the 32-bit data storing into the register file; this information allows the stored data to be verified later on.

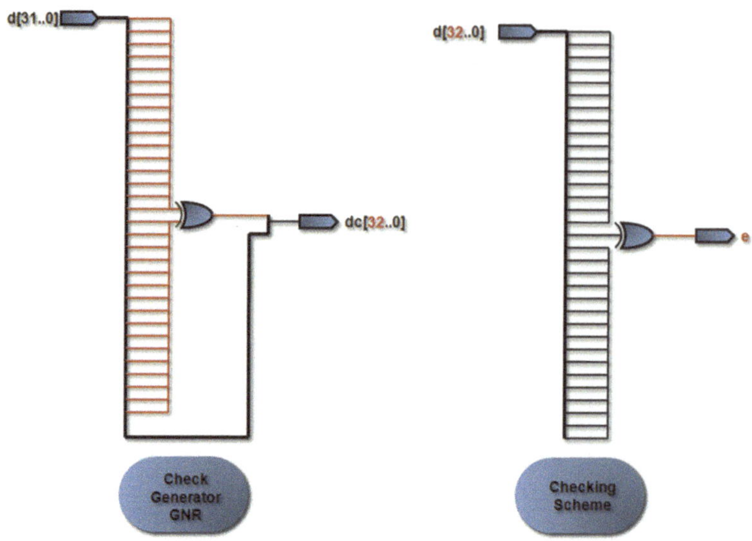

Fig. 6.4 Check generator and checking scheme

The Checking schemes (marked in blue in Figs. 6.3 and 6.4) check the data integrity when data are read out from the Register File and when it is possible; correct the data before the ALU operation takes place.

In order to implement the fault recovery process P2 an extra Register Buffer Rbuf(R*) is introduced (marked in red in Fig. 6.3). The register buffer is allocated to keep premodified state of operand for the currently executed instruction.

When a fault is detected during instruction execution, it allows the processor to restore to the initial state before the execution of the instruction, enabling the instruction to be repeated. This enables us to tolerate faults within instruction execution.

The extra Register Rbuf(R*), the checking schemes and the reverse instruction sequencer combine to make the implementation of P1 and P2 possible without any perceptible time overheads.

6.2.2 *Passive Zone*

Figure 6.5 presents the current custom hardware prototype of the ERA device. The prototype is provided with two flash based ROM modules 128 Mb each (8 Mb $\times$ 16 bit) with replicated bootstrapping firmware and operating system software.

The proposed memory scheme may be regarded as a collection of four blocks, 16-bit wide, with identical size of 1 Mb each (16 $\times$ 64 k). Using 16-bit memory modules instead of 32-bit memory modules increases reliability and reduces, when

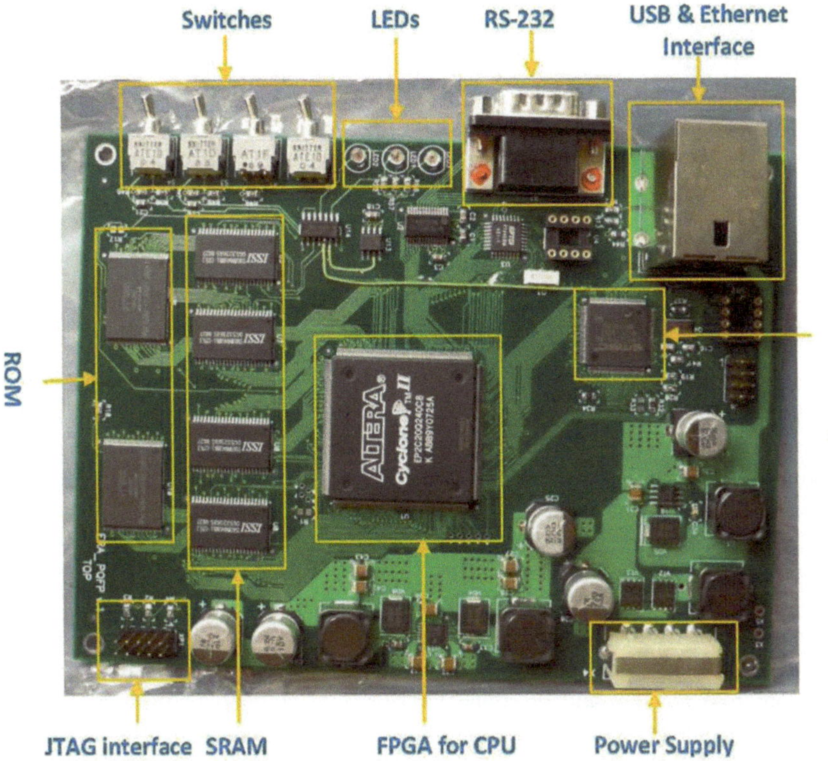

Fig. 6.5 ERA prototype board

necessary, energy required for execution. Additionally, the scheme includes two flash-based ROM modules with replicated bootstrapping firmware and operating system software.

Reliability is increased by means of added working states and configurations.

Energy-wise operation is improving by means of this architecture ability to activate only modules required - by means of using a single 16-bit memory module, when necessary.

The proposed memory scheme allows different configuration schemes, explained in Sect. 6.4.3.

6.2.3 *Interfacing Zone*

The above explained architecture principles of design relate to the interfacing zone as well. The architecture needs something (logic) in the interfacing area to enhance the flexibility and resilience of system operation between active and passive zones, for wide range of applications where PRE-properties are key requirements.

This motivates the ability to have a reconfigurable interfacing zone. One of schemes that we propose is known as T-logic. T-logic is a hardware element that provides reconfigurability of the architecture for performance-, reliability- and energy-wise operation. This logic should be able to provide minimal configuration—when one processor and memory remain.

Due to importance of T-logic as "major executive agent" for reconfigurability of the whole system next section explain it in details.

6.3 ERA Reconfigurability

As declared earlier performance, reliability and energy-awareness are required for the next generation of computer systems. Reconfigurability requires hardware and system software implementation and support. In order to be able to change the configuration when necessary (sometimes several times during a single mission) systems should have special elements with specific properties such as extreme reliability, performance and simplicity, supported by independence from faults of the system.

For the purposes declared above we propose a hardware element called T-logic. The main function of a T-element is to connect and disconnect the system component, using "logic rotation" for various types of connections and configurations.

In order to connect and disconnect the system component there are several "turns" of T-element as Table 6.1 below shows, describing the basics of T-Logic element.

The reconfigurable interconnection schemes can execute dynamic reconfiguration of system transparently from for software. For example, when we use triple system elements configuration, T-elements might exclude faulty ones from operation, leaving only two elements active (DMR). If further degradation is allowed the configuration (in terms of fault tolerance) can continue up to a single element (Schagaev and Buhanova 2001).

Figure 6.6 below illustrates how reconfiguration might be used for reliability purposes using T-elements. The figure shows a diagram that exemplifies a scheme with three memory modules working in parallel that at some point during operation experiences two permanent failures in two out of three modules.

6.3.1 T-Logic for Memory Management

ERA power consumption can also be controlled using T-elements. Existing electronic technologies posses the following drawback: increase in power consumption causes degradation of system reliability. Therefore, an ability to connect and disconnect a system element might be function and requirement in real time and other applications.

Table 6.1 T-logic rotation

Position+	Description+
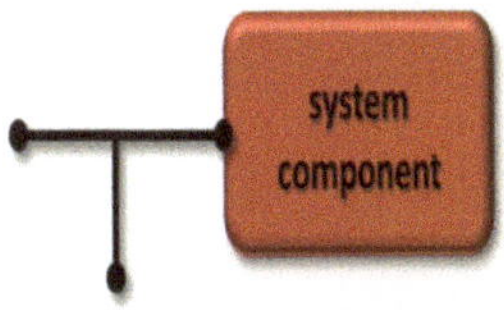	The “T” logic connects the active zone to a front element that is connected in redundant mode with a right element. This front element leads the rest of the elements that is connected to
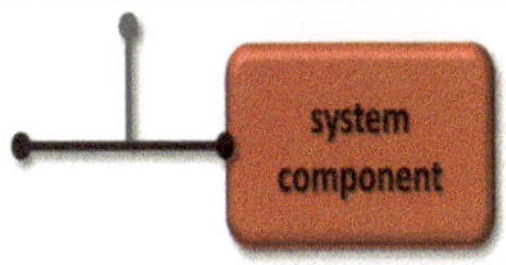	The “T?logic” connects the active zone to a front element. In this case the element is working in serial mode
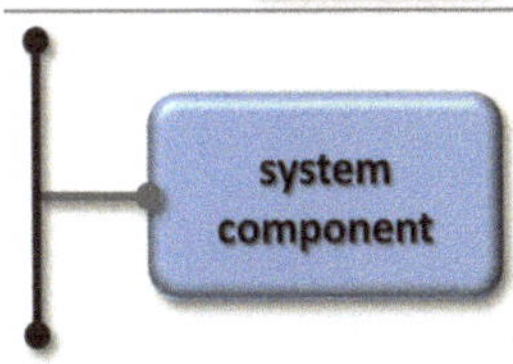	The “T?logic” connects the element to two neighbouring elements in redundant mode. The element is leaded by the left side component that is connected to the active zone
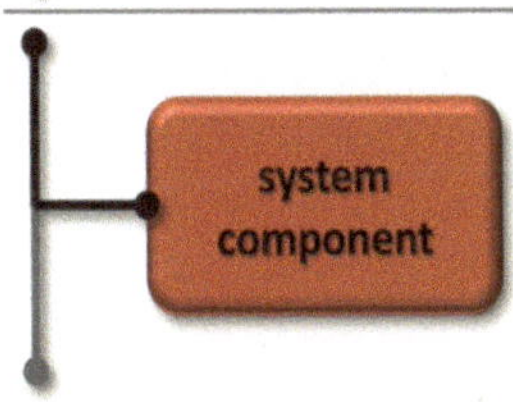	The “T?logic” connects the element to a left element. system component will be leaded by side element that is connected to the active zone
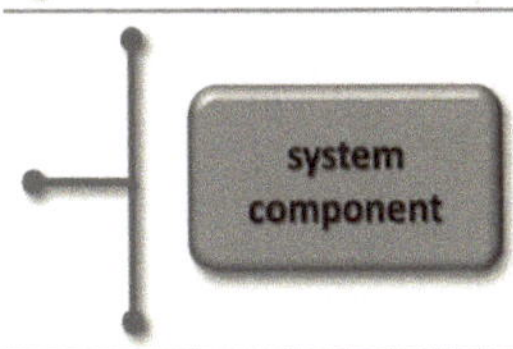	Disconnected the element: The “T?logic” is disconnected from the interconnection scheme. The energy consumption of the element is reduced to the minimum

Again, for illustrative purpose we use an example when ERA uses three system elements in redundant mode. If the task scheduled does not require a full size configuration, the architecture can be configured to operate using either two elements or a single element on its own.

Hardware configuration should be implemented transparently to application programs. Task defined requirements might be available for a run-time system. Excluding memories will not change the logic of the program.

Figure 6.7 shows the power saving algorithm for a reconfigurable system. The reconfigurable interconnection schemes improve the memory management flexibility. For instance, if the computer architecture has three elements and the task requires maximum capacity, the system will configure all memories in redundant mode so that it can provide maximum capacity.

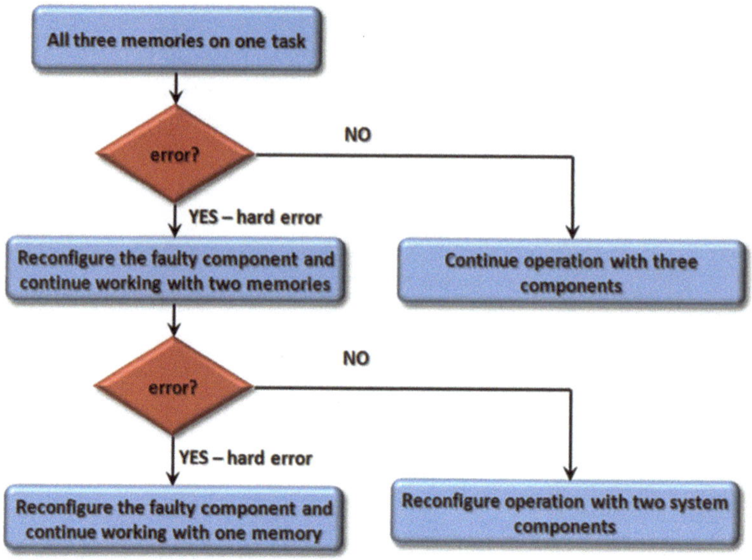

Fig. 6.6 Algorithm of configuration for reliability using T-logic

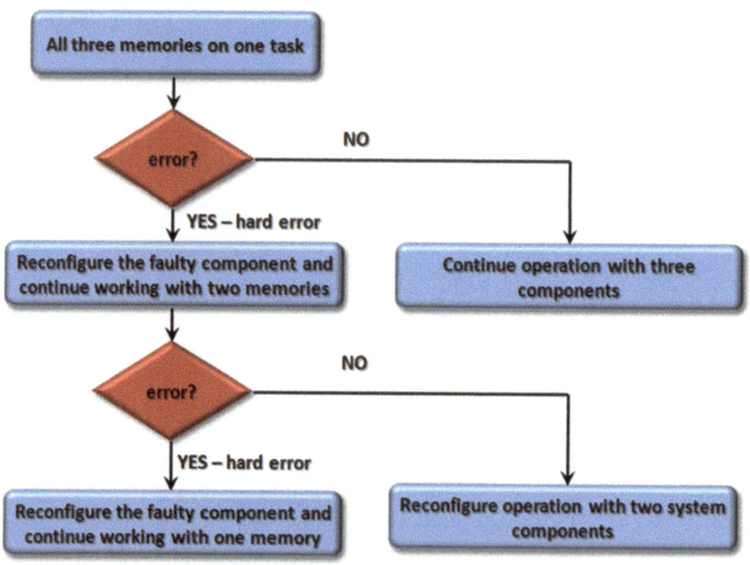

Fig. 6.7 Energy-wise algorithm of configuration using T-logic

6.3.2 *T-Logic for Configuration in ERA*

Possible configurations of a system that uses "T" logic are presented in Table 6.2 below. The first Raw illustrates maximum reliability configuration (if three HW elements are available). The power saving of the system could be improved by disconnecting elements from system and keep them idle—raw 2.

The configuration for maximum capacity required for application is shown in the third raw.

To provide fail-safe functioning for ERA T-logic itself should provide the following property:

- when "T" element itself fails system functioning should continue. Its design, therefore, must implement fail-safe principle; in this case T logic element should be replaced by wire to enable further system functioning. This all in fact defines requirements for interconnection zone for ERA PRE-smart functioning.

Table 6.2 Possible system configurations using T-logic

Configuration	Explanation
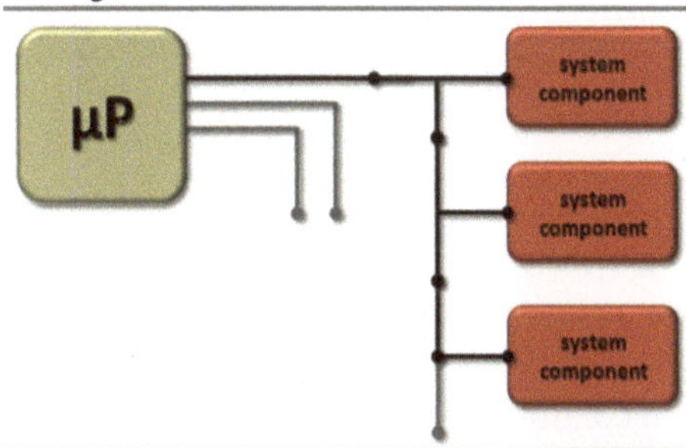	"T" configurators connect all three components with processor. Top system component acts as leading element. The rest system elements compare the results and participate in voting. Thus reliability of this system configuration is high. "T" element configures enables excluding faulty system component
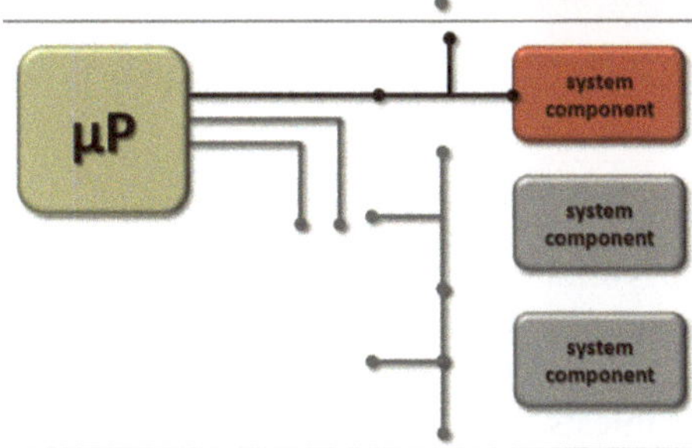	This system configuration serves for maximum energy saving. In this case "T" element connects only one system component with processor, while the rest are idle
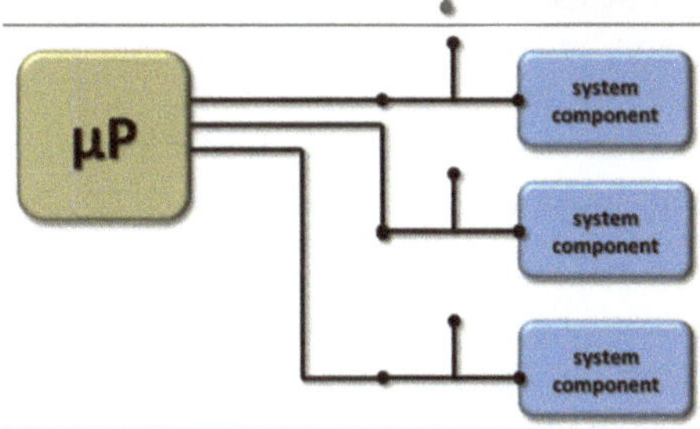	In this case, all three components are used for maximum hardware capacity. When performance of application is main priority this configuration fits the purpose

This illustrates how principle of separation of concerns can be applied for design of hardware schemes.

6.4 Syndrome

As mentioned earlier, the system new property must be supported by hardware and system software implementation of required processes that make this property. We introduce for this purpose a special hardware scheme called *a syndrome*. The term *Syndrome* is new Latin (origin 1535-45) and was originated from Greek "syn-drome" where:

- "Syn-" from combination, concurrence.
- "Dramein", main meaning is "to run".

For our purposes a syndrome is not passive, presenting "a snapshot status" of a system, *a Syndrome* also is active, a serving tool to control the system configuration.

Thus, a *syndrome* is "*a group of related or coincident things, events, actions, signs and symptoms that characterize a particular abnormal condition*".

Further analysis and development of syndrome concept and application follows.

6.4.1 Syndrome Use

Clear that functions of a syndrome are not only passive, presenting "snapshot-status", but also active, serving as a tool to control of the system configuration and enabling to estimate system conditions.

Syndrome also might help to answer a question that is shamefully omitted in the vast majority of research about fault tolerance:

WHAT PROVIDES THE FAULT TOLERANCE OF THE SYSTEM?!

It is usually assumed that the core logic is ultra reliable and guarantees control of configuration and reconfiguration. Unfortunately, using the homogeneous redundancy may limit the increase in reliability since techniques based on the same type of redundancy are vulnerable to the same threats.

Hybrid techniques based on heterogeneous redundancy can be more effective.

Thus, even when checking schemes of memory or configurators or processor detect error and transfer information about it to the syndrome, this information might not be useful if the system does not include either one or both mentioned below:

- "External elements" responsible for exercising GAFT and making decision on configuration/reconfiguration if necessary.
- Internal element that is capable to initiate required sequence.

Indeed, in regular computing system when there are faults in the processing logic, to expect to expect that it is able to perform self-healing and then control and monitor configuration of the rest of the system is an attribute of fairy tale, not engineering and seems to be unrealistic. There is a solution though, as described below.

Figure 6.8 shows conceptually ERA active zone divided by two AU and LU elements.

To be able to absorb and trust an information about a status of an element, every checking signals about condition of registers (not shown) memory, AU and LU as well as control unit should be aggregated in syndrome.

The scheme of implementation of fault tolerance separates the passive zone and active zone of the proposed architecture. A clear separation of the functions of processing (of data operation) and storing (memory) enables to apply various checking and recovery solutions. The passive zone has the elements to store data, while active zone is for the data manipulation.

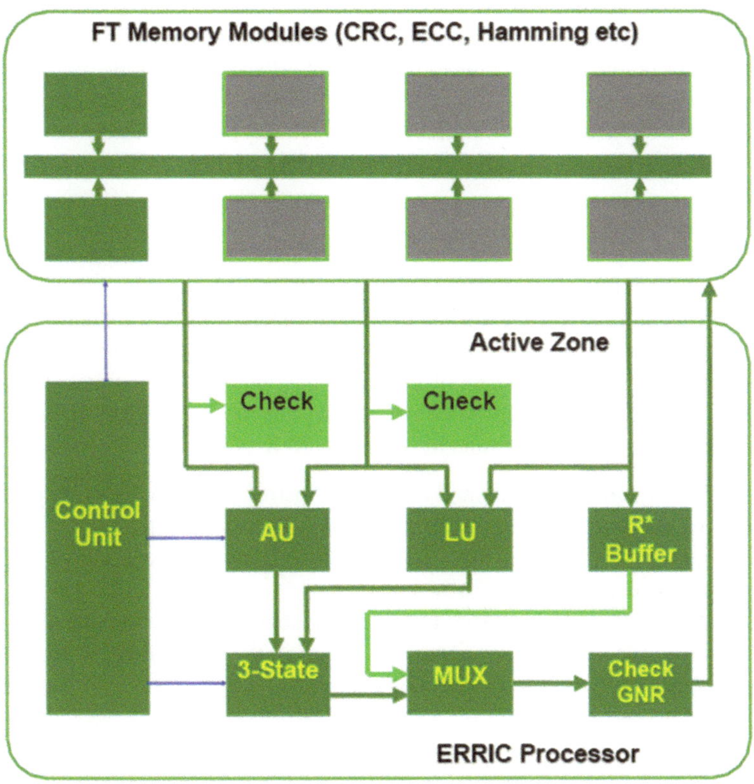

Fig. 6.8 Processor-memory structures with “separation of concern”

All processor registers (register file) may be protected by parity or other checking schemes. During instruction execution data are loaded from the register file into functional elements for operation and operands are checked.

If there is no fault detected during instruction execution the operation is considered as successful and the result is stored back into the register file or sent out to memory. The fast and reliable memory access is possible by application of static memory and register files.

The memory context data might be protected by schemes such as error checking code, Hamming code, or recently proposed schemes (Gössel et al. 2008).

If error is detected, the control unit by executing GAFT attempts to restore the damaged data and repeat the instructions:

- Restore the damaged data.
- Repeat the execution of the instruction that manifested a fault.
- Resumed execution.

ERA provides the fast and reliable recovery scheme within instruction execution level, transparently for software.

Taking into account requirement of minimisation of redundancy the architecture presented in Fig. 6.8 above seems to be efficient—there is no duplication or greater level of reservation applied for active zone. Active zone consists of two non-identical units: arithmetic and logic units respectively, (denote them AU and LU).

In terms of power consumption this scheme is also efficient. A question 2 arises:

IS IT POSSIBLE TO TOLERATE FAULTS WITHOUT DUPLICATION?

The previous question becomes crucial for implementation of fault tolerance for any system and proposed scheme with minimum redundancy. The solution is possible if we apply well-known mathematical results about equivalence of arithmetic polynomial for logic functions (Vykhovanets 2006), attempting to represent arithmetic functions by sequence of Boolean functions.

There is a theory when arithmetic functions are represented by Boolean functions. These two groups of results hint us a possible solution:

- An ISA that consist of logic and arithmetic instructions should be supported by sequence of functional equivalence of logic operators implemented by arithmetic unit. For every arithmetic instruction a functionally equivalent sequence of logic instructions should be added.
- Both sequences should be stored in different segments of read-only memory.
- If a fault occur in the AU a signal sets a flag in the syndrome and sequence of logic operators is executed instead of arithmetic instructions to complete GAFT, and vice versa.

Note that this equivalence and hardware redundancy is used only when a fault is detected and the recovery procedure is initiated.

This enables the system to recover from transient errors and execute reconfiguration. After reconfiguration, the system can either continue functioning in normal mode or, in case of an unrecoverable permanent fault, it can provide fail stop sequence of actions.

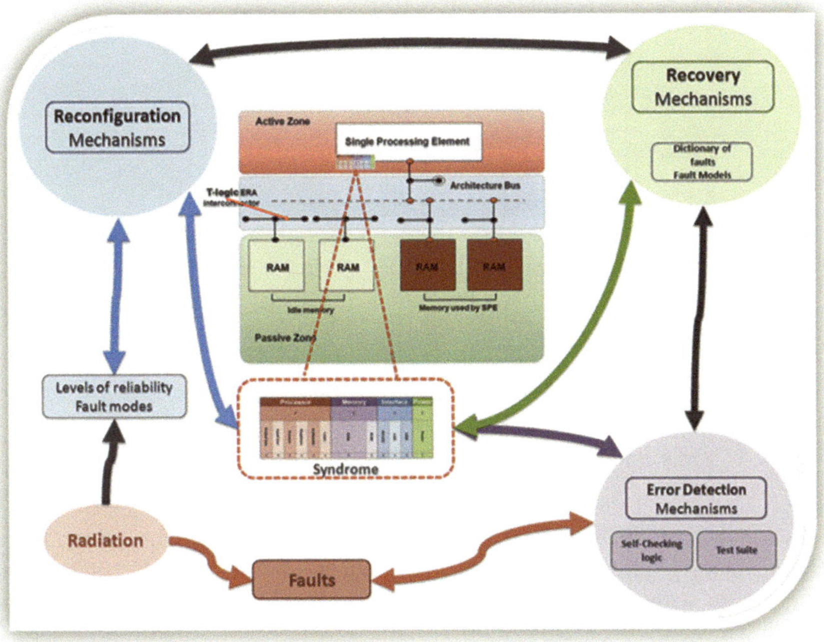

Fig. 6.9 Syndrome purposes

Area of applications for embedded safety critical systems set hard constrains for instructions execution and replacement of arithmetic instructions by sequence of logic ones has no justification to apply for execution of application programs.

The syndrome acts as a control centre for three main functions: fault monitoring, reconfigurability and recovery (Fig. 6.9). These three functions serve for the purpose of performance, reliability and power efficiency.

From a hardware point of view the syndrome is represented as a special hardware register that will interact with the system via hardware interruptions schemes.

Semantically, the structure of the syndrome is subdivided in three different areas (Fig. 6.10):

- Fault control area.
- Configuration control area and,
- Power control area.

The Fault control area reflects the hardware status of the different areas of a single processing element: processor, memory and interface. Full "Zero" syndrome in this area indicates that no fault has been detected in the system. If a fault in a specific element is detected, the corresponding bit is set to 1.

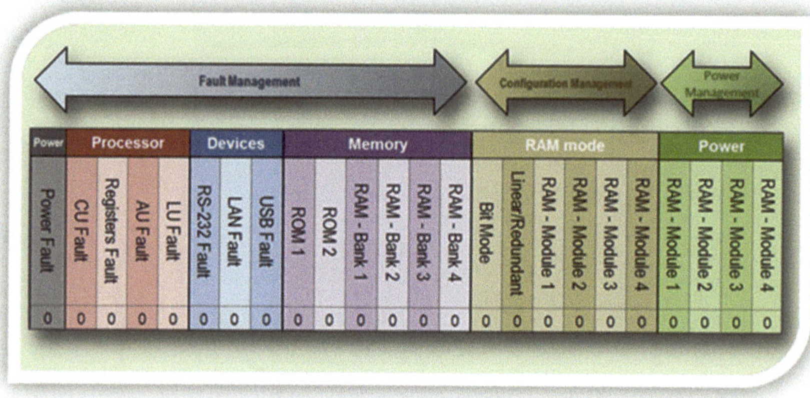

Fig. 6.10 Syndrome for reconfigurable architecture

Each hardware element has sits own representative flag within syndrome. For ROM memory group that consists of two chips syndrome of ROM condition has two positions with zero when ROM works correctly. In turn, static RAM memory has four hardware elements.

The configuration area of syndrome reflects memory mode that ERA is currently using. The Bit mode field defines whether the addressing mode is 16 or 32 bits, whereas the L/R field defines whether the memory banks are in linear or redundant mode.

Bit mode:	0 = 16 bits
	1 = 32-bit addressing mode
LR:	0 = Linear
	1 = Redundant

The power management area reflects the status of the modules in terms of power.

Bits Power RAM Module 1, 2, 3 and 4 represent whether the Memory module is powered:

0 = Power Off
1 = Power On

The combination of those three areas: Fault, configuration and Power management control defines the state of the system. For example, a memory module could be in the following states: faulty, failed, stand-by, ideal and off state. Checking those three areas by a simple reading would determine the status of the memory module. When reconfiguration is set by software, the states of syndrome might be mirrored in memory. Keeping those states only in SSW is not universal solution, for example when external element requires an access to a syndrome and local memory is failed further use of available resources and elements becomes impossible.

Without a doubt, the syndrome is one the most critical parts of the system. For reliability purposes, there are three copies of the 32-bit register syndrome connected to a voter within the processing element. Triplication of the syndrome increases complexity of logic. The voter will be vulnerable as well (unless the voters are triplicated). Another option that solves the complexity would be low-level hardening techniques and/or using different technologies (such flash memory) just for the syndrome register.

However, that will increase the manufacturing costs. Without this replication, and possibly different, enhanced technology use a bit flip in the faults area of the syndrome would lead to redundant fault detection processing, whereas a bit flip in the configuration area would likely end up in a catastrophic failure.

6.4.2 *Location Access and Way of Operation of the Syndrome*

There are two major mechanisms that will be able to detect a fault: hardware logic and SSW. Both, hardware logic (mismatch and voters) directly and SSW (testing and detection mechanisms) via instruction should have access to the syndrome:

SSW events: If SSW has access to the syndrome, reading and writing the value of the syndrome is compulsory. Note the current ISA does not include instructions within the processor to do that, unless one of the registers (e.g.: register 31) is used as a syndrome. This is similar to the VAX V70 processor [Kimura88] that includes a status word placed in a fixed address of the memory. In this case, the register file needs to be hardened. At least the syndrome needs to be hardened.

A bit flip in the configuration area of the syndrome could mistakenly turn off one of the memory modules. As a consequence, a preferred way to access the syndrome that avoids changing the ISA and hardening the register file is the use of the input/output memory lines (mapped in a reserved address). This works as a TMR scheme, with complete software independence.

Regardless of the hardware implementation, only one syndrome is visible to the system software. An error in one of the syndrome registers is corrected by the hardware without the software intervention.

However, it is useful that the SSW is aware of errors in the syndrome. Errors within the syndromes are useful information in a potential contingency plan (e.g.: setting the fault tolerance of the system to a higher level in case of recent particle impacts).

6.4.2.1 Automatic Events Detection Mechanisms Using Hardware

Special HW interruptions are needed for this; if during the diagnosis of a memory chip the ALU for example signals a problem, a diagnosis of the suspected element should be done first.

The syndrome might be considered as independent new hardware—independent from processor hardware elements and other schemes.

A method is needed for synchronising the operation of the processor with the syndrome. One solution would be polling, where a loop that checks the status of the syndrome is arranged. It has a major disadvantage: the processor is busy reading the syndrome, instead of executing some useful code (wasteful in terms of processing power).

Instead of polling the syndrome waiting for a change, a hardware interruption system is preferred.

In this case, the syndrome subsystem is responsible for notifying its current state to the processor.

When the syndrome needs the processor's attention, it sends an electrical signal through a dedicated pin in the interrupt controller. In this case, the processor stops its current activity and jumps to execute a function (interrupt handler) which has to be associated with the fault manifestation.

By using hardware interruptions, in terms of total execution time, the syndrome will be accessed only when a fault is manifested and reflected in the syndrome. Most of the time the fault area of the syndrome will be 0 and the rest of the areas will only be altered when restarting of the system or when changing the memory mode.

The same method might be applied to the control of other devices. Different levels of interruptions are then needed. Active zone hardware should have higher priority than passive zone hardware.

If a hardware mechanism sets the syndrome bits, then the processor executes a trap (exception) and starts the diagnosis software.

The software treats the fault and clears the syndrome otherwise the processor would trap again and restart the diagnosis.

This scheme introduces a requirements to set and reset syndrome register internally (using own hardware or software) or externally, from the "rest of the system" when ERA is used in the form of CC—connected computer structure.

If the syndrome is located within the active area, in the case of a faulty active area a neighbour processing element may have difficulties accessing to it. To resolve it there are three options:

- To enable system flexibility in fault handling, one has to implement syndrome independently from the active area where it can be accessed via hardware by neighbour single elements.
- Software message passing: Instead of hardware access SSW will deal with status of single elements, by sending un update (periodic updates) on syndrome to the single element neighbours before changing memory mode or when a fault has been detected (if feasible in this last case). A syndrome table in a similar fashion (but not as heavy) to the routing tables used by routers and the different routing algorithms could be shared by different processing elements.
- Both, Hardware access and SW message passing are not mutually exclusive. A combination of both is possible.

6.4.3 Syndrome: Passive Zone Configurations

A total of 25 operating states of memory are possible to operate in reliability-, energy- or performance-wise modes. The characteristics of the proposed memory architecture are given in Tables 6.3 and 6.4.

Memory modes are subdivided into two major categories, depending on the number of bits read/written at once: 32 bits and 16 bits. At the moment there are ten different usable configurations in 32-bit mode memory (defined in Table 6.4) and 15 different configurations in 16-bit mode memories (defined in Table 6.3). In addition, two modes of operation, depending on the existing amount of redundancy, can be selected:

Table 6.3 16-bit addressing modes in RA

		RAM mode		Information - Power								
State	Phase	Bit Mode	Linear/Redundant	RAM - Module 1	RAM - Module 2	RAM - Module 3	RAM - Module 4	Space (Mb)	Reliability	Speed	Power Consumption	Syndrome
11	4	16	L	A				1	1	2.5	2	111000
12	4	16	L		A			1	1	2.5	2	110100
13	4	16	L			A		1	1	2.5	2	110010
14	4	16	L				A	1	1	2.5	2	110001

Table 6.4 32-bit addressing modes in RA

		RAM mode		Information - Power								
State	Phase	Bit Mode	Linear/Redundant	RAM - Module 1	RAM - Module 2	RAM - Module 3	RAM - Module 4	Space (Mb)	Reliability	Speed	Power Consumption	Syndrome
1	1	32	R	A	B	A	B	2	2	4	4	111111
10	3	32	L	A	B	C	D	4	1	5	4	101111
2	2	32	R	A	B	A		2	1.5	4	3	111110
3	2	32	R	A	B		B	2	1.5	4	3	
4	2	32	R	A		A	B	2	1.5	4	3	111011
5	2	32	R		B	A	B	2	1.5	4	3	110111
6	3	32	R	A	B			2	1	5	2	111100
9	3	32	R			A	B	2	1	5	2	110011
8	3	32	R	A			B	2	1	5	2	111001
7	3	32	R		B	A		2	1	5	2	110110

1. Linear mode: where no module is replicated and
2. Redundant mode: with at least one module being replicated.

In order to explain the available memory modes, let us consider the letters in the set {A,B,C,D, x} as a representation of information stored in the 16-bit memory modules, where:

- A lowercase letter x represents a module is not in use.
- Identical letters represent identical information in different modules. In other words, information stored in a module is n times replicated into n modules: e.g. AA, AAA, BB and AAAA. In this case n-1 modules are connected in shadow mode and perform all memory operations concurrently to their respective master memory module. A hardware voter in the memory controller compares the output of the memory modules and triggers a fault in the syndrome in case of mismatch.
- The pairs AB, BA and CD represent two 16-bit modules combined into a virtual 32-bit module.

6.4.3.1 32-Bit Mode

The memory modules of ERA are 16-bit-wide. Therefore, two memory chips must be combined to allow 32-bit memory access. Table 6.4 reflects all the supported memory combinations in 32-bit mode.

Only one memory mode is available in 32-bit Linear Mode (state 10 in Table 6.4). Module 1 and 2 are combined and mapped in the memory space. Module 3 and 4 are combined as well and mapped contiguously in the memory. This configuration provides maximum space for the application but does not feature fault tolerance, and not even fault detection at the memory level.

In terms of fault tolerance, the maximum redundancy based on 32-bit addressing is the ABAB configuration (Table 6.4). The information in the two modules AB is replicated on an extra pair AB. In every reading operation the data from both memory pairs is compared. If there is a mismatch, the checkpoint area in RAM will be the decisive factor in selecting the failure.

Note that 32-bit modes that involve three working modules such as ABAx ABxB, AxAB or xBAB are not initial modes (starting in this mode does not offer any advantage). Those four modes would typically involve that an error has occurred in one of the modules ongoing diagnostics is taking place.

6.4.3.2 16-Bit Mode

With the main purpose of critical energy saving a 16-bit addressing mode was introduced. This mode is using a single bank of memory. Hence, there is not duplication of memory. The four different memory configurations, currently allowed in this mode are presented in Table 6.3.

When only one single mode is available due to permanent faults in the other three, the system could be restarted in a fresh 16-bit mode loading the different SSW binary codes from a ROM location into RAM. Note that this mode is an emergency mode and does not contemplate the possibility of hot switching from a 32-bit mode. Schemes to change from 32-bit mode to 16-bit/power saving mode and vice versa are defined by the runtime system.

6.5 Graceful Degradation

If one of the memory modules fails, the system can be reconfigured to exclude it from the current configuration. In terms of fault tolerance, the maximum redundancy based on 32-bit addressing is the ABAB configuration (Fig. 6.11).

The ten admissible states in 32-bit mode are given in Fig. 6.11. Further degradation is possible in 16-bit mode (Fig. 6.12). In terms of fault tolerance, the maximum redundant configuration is based on 32-bit addressing ABAB configuration. The transition between these states is completely dependent on soft/hard errors and the efficiency of recovery mechanisms. However, a voluntary transition between different states is also allowed.

In case of degradation, successful recovery mechanism could produce a transition from a degraded phase to an initial phase. The group of states or phases can be distinguished:

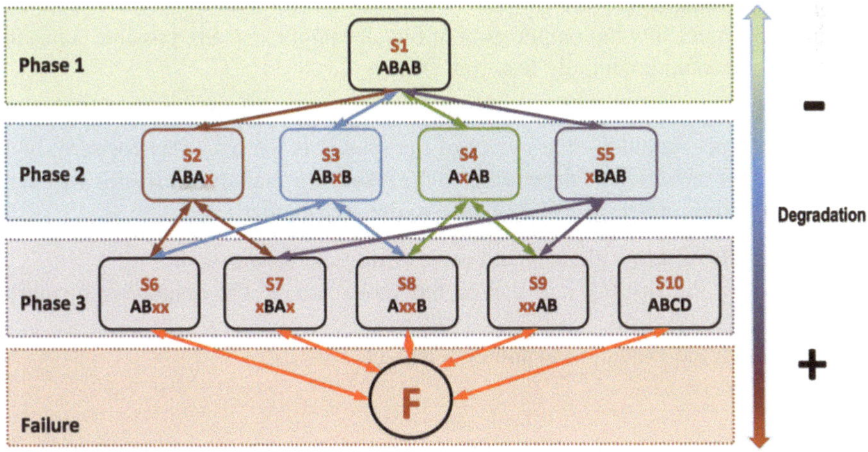

Fig. 6.11 32-bit degradation phases

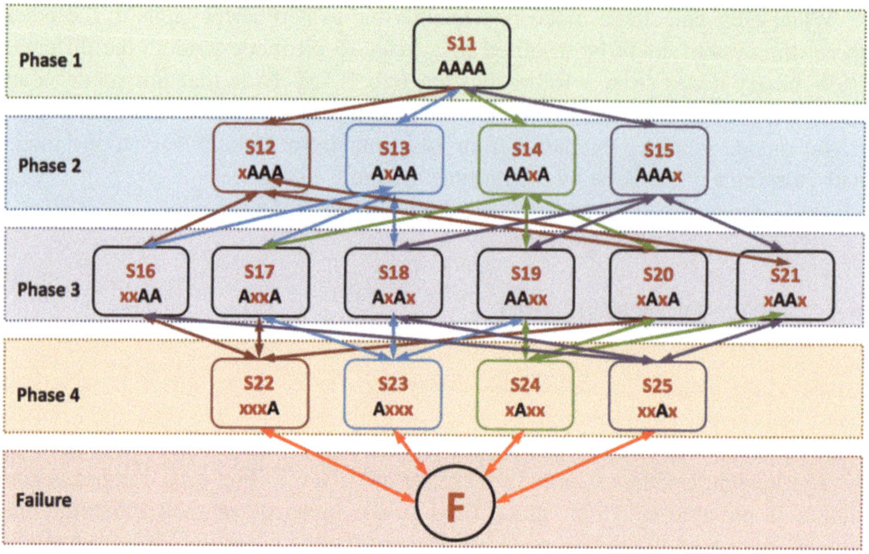

Fig. 6.12 16-bit degradation phases

- Phase 1: States with full checking and replication of every single bit. The single state in this phase is the initial state and has maximum redundancy for 32-bit ABAB configuration (see Table 6.4, Fig. 6.11).
- Phase 2: States in which at least 50 % of the bits are replicated. Transition to one of the four different states available in this phase is due to a fault in one of the memory modules of State 1.
- Phase 3: States in which replication of bits does not exist. Six possible states in this phase could potentially lead to a failure.

If during operation phase 3, working module experiences a third permanent error (at least one per module), then reboot of the system is the only way forward. In a multi-element scenario and depending on the resources available, reliability needed and current environment, SSW will have to decide whether to:

- cold switch the same element and restart using the 16-bit mode.
- keep using the healthy active area but make use of the neighbour memory elements or,
- cold switch the execution to another element.

6.5.1 *Graceful Degradation: Markov Analysis*

Regarding computational capability and availability of resources, a system can be modelled as being in one of many possible states. The number of states would be large, if fine distinctions are made, or it may be relatively small if similar states are grouped together. Different events can force the system moves from one state to another depending on resource availability and computational changes.

By quantifying the probability of state transitions, State-space modelling can determine the probability of the system being in each specific state; this can be used to obtain some parameters of resilience (reliability, safety, maintainability, safety, etc.).

The sum of all input and output transition probabilities of each state should be 1. The state of the system is characterised by the vector: (S_0, S_1, S_2, S_3, …, S_n).

A transition probability matrix has N states. On the t'th time-step the system is in exactly one of the available states q_t:

$$q_t \in \{S_1, S_2, S_3, \ldots, S_n\}$$

There are discrete time-steps, $t = 0$, $t = 1$, …. We are interested on how the system will behave after several time steps. Initial condition of the system: S_0. Given q_t, q_{t+1} are conditionally independent of $\{q_{t-1}, q_{t-2}, \ldots, q_1, q_0\}$ that is:

$$P\left(q_{t+1} = s_j \middle| q_t = s_i\right) = P\left(q_{t+1} = s_j \middle| q_t = s_i, \text{ any earlier history}\right)$$

The current state q_t determines the probability distribution for the next state q_{t+1}. In order to model of the system we do the following assumptions:

- System starts in the perfect state.
- Only one fault can occur at a time.

In order to simplify we make a first-order Markov assumption: we say that the probability of an observation at time n only depends on the observation at time n only depends on the observation at time $n - 1$. In a sequence $\{q_1, q_2, \ldots, q_n\}$

$$P\left(q_n \middle| q_{t-1}, q_{t-2}, \ldots, q_1, q_0\right) \approx P\left(q_t \middle| q_{t-1}\right)$$

Using the previous assumption, the joint probability can be expressed as:

$$P(q_1, q_2, \ldots, q_n) = \prod_{i=1}^{n} P\left(q_i \middle| q_{i-1}\right)$$

Figure 6.13 presents Markov model of reliability for reconfigurable memory described above. The figure shows the transition probability in the case of four-module memory scheme with TMR plus a spare configuration. The circles in the figure represent one of the 15 possible states. The arrows reflect transitions from source to destination state.

By merging states in the Markov model a simpler equivalent model is created. Figure 6.14 represents such simplification illustrating the probabilities of transition between the original TMR plus spare, a TMR, a DMR, a SMR and Fail states.

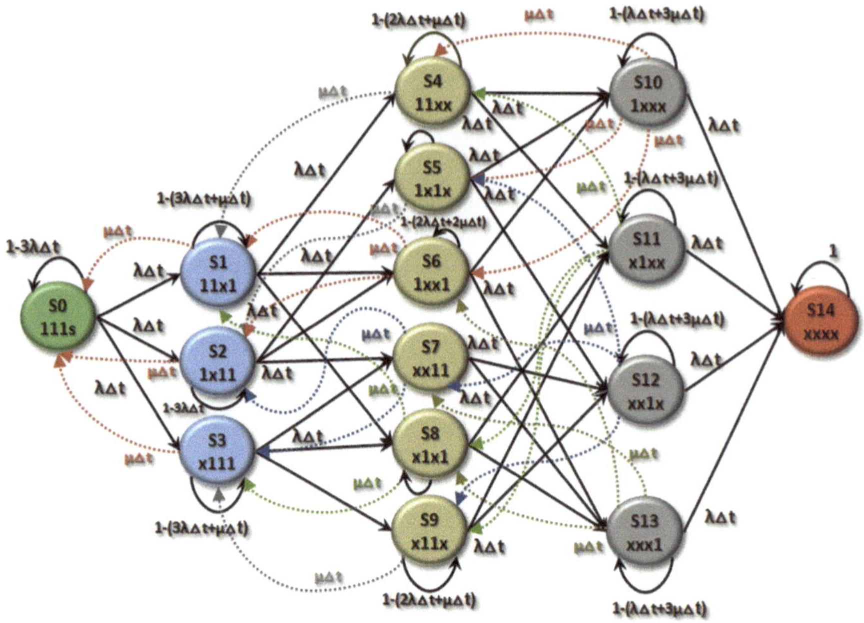

Fig. 6.13 Markov model for the ERA memory system

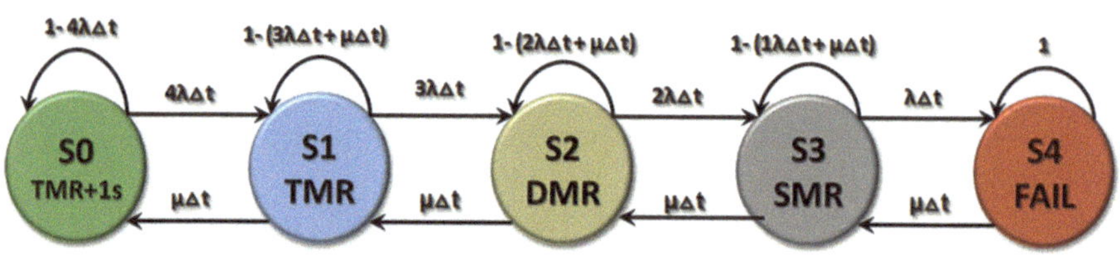

Fig. 6.14 Reduced Markov model for the ERA memory system

6.6 Implementation Constraints

As explained before in Sect. 6.2.1 taking into account the 16-bit instruction size and 32-bit word size of memory organisation, at the time of compilation, the compiler schedules memory loads on the first 16-bit instruction of a 32-bit two-instruction "packet". This way, memory loads and instruction fetches never occur at the same time. When the memory configuration is set as 16-bit words each read or write precede instruction execution. In the case of a 32-bit memory configuration two instructions might be loaded from memory by one access.

Memory Configuration																															
31	30	29	28	27	26	25	24	23	22	21	20	19	18	17	16	15	14	13	12	11	10	9	8	7	6	5	4	3	2	1	0
																											1	M1		M2	

Bits	Name	Meaning
[0:1]	M2	The number of the memory module used for the higher 16 bit memory
[1:2]	M1	The number of the memory module used for the lower 16 bit memory

Fig. 6.15 Theoretical memory configuration with support of reconfigurability

6.6.1 Graceful Degradation: Markov Analysis

Different solutions could have been implemented for ERA's memory addressing scheme. We consider physical addressing and relative addressing.

Following the first case, as shown in Fig. 6.15, the lowest 4 bits of 32-bit memory address would represent which modules are used to represent the lower and the higher 16 bits. The encoding would be as follows: Bit 2 and 3 correspond to M2 and represent the higher 16 bits.

However, by using this scheme, having the configuration of the modules physically mapped in program code would make reconfiguration difficult to implement. If binary code reflects which bank the program should run on, then in case of fault that requires code to be transferred to another bank (bank switching), a translation would be needed.

This recompiling of the program is highly unlikely possible in real time of program runs. This also affects recovery time and, in fact, it excludes the chances of recovery in real time.

Besides, if the contents of the memory banks are physically different this will affect the hardware checkers complexity; the checker function would need to compare equal values in case of data comparison and different values in case of address comparison. Thus, separation of concerns principle and a virtual memory approach are preferred.

By removing M1 and M2 (Fig. 6.15), the memory addresses used in a program code refer to a relative position, for instance, within a pair of modules AB. A reconfigurable memory controller (see Fig. 6.16) that links the relative address to the specific memory bank is used in a similar fashion to the translation of virtual addresses into physical addresses.

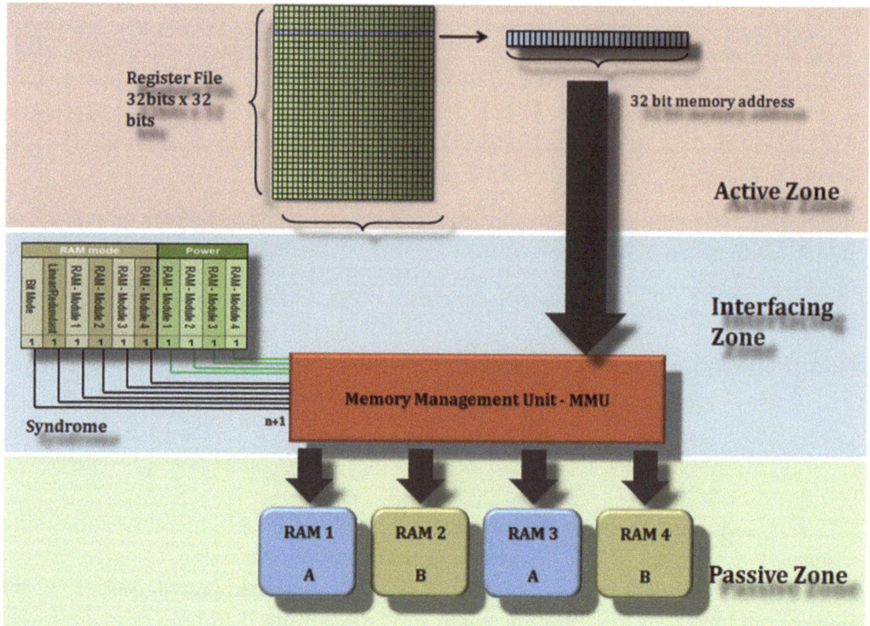

Fig. 6.16 MMU syndrome as a memory configurator

6.6.2 *Interfacing Zone: the Syndrome as Memory Controller*

As previously seen in Table 6.1 the T-logic ERA element performs configuration and reconfiguration of hardware by providing interconnection and dynamically excluding faulty components from the operational system.

The T-logic interconnector provides flexibility of application of memory elements (32- and 16-bit configurations) and at the same time helps in fault containment. This logic is used to form a hardware configuration scheme adjustable to the program requirements or when a hardware element itself (or architecture) detects hardware faults and thus can't be involved in further calculations.

Note that "isolation" might be temporary or permanent, subject to the element's "health". The final decision about permanent isolation of an element will take place when testing and recovery procedures are complete.

The four T-logic interconnectors, one for each memory bank, are physically included in the T-logic Management Unit or TLMU. TLMU (see MMU in Fig. 6.16) manages the connectivity of the memory, configures and reconfigures the working mode to a 16-bit single memory, 32-bit double memory with master/slave configuration or any of the 14 memory addressing schemes available (Tables 6.3 and 6.4). Using the T-logic scheme memory elements could be isolated, switched off for power reduction or doubled in capacity when the maximum storage volume is required, addressing PRE-wise computing.

The Configuration and Power management flags of the syndrome describe the different states of the memory modules. Different values in the configuration area of the syndrome select the bank used and the mode.

The output memory lines of the processor determine a location within a memory bank, whereas the Configuration and Power areas of the syndrome specify which banks are to be used and in which mode.

One example of a possible memory configuration (State 1; Table 6.4) arranged by "T"-logic is presented in Fig. 6.16. The example reflects a 32-bit (Bit mode = 1) ABAB configuration with 2 × 2 modules duplicated (Redundant = 1) working in pairs.

By using this method we can increase the independence of software/hardware configurations for the PRE-purposes. Memory addresses within the code do not need to be arranged, as code integrity is a crucial requirement for safety critical systems.

Let us define the following scenario where it is required to switch data and code from modules 1 and 2 to module 3 and 4. Let us assume that the work mode is ABxx 32-bit, which is fast but not very reliable.

Assuming that the user (or an online fault detector manager using a fault model) requires a higher level of reliability, transfer from the current ABxx mode to an ABAB mode is required. An implementation of this example is based on the following algorithm (omitting the testing procedures):

```
!i!=!0;!
REPEAT!
!   i++;!
!   Starting!at!memory!location![0]!load!the!n!following!words!into!RF!(32x32)!
!   Change!the!value!of!the!syndrome!to!the!memory!mode!XXAB!
    Copy!the!n!previously!stored!words!from!the!RF!starting!at![0]!+!i*32*32!
!   i++!
!   Change!the!value!of!the!syndrome!to!the!memory!mode!ABXX!
UNTIL!EOM!(End!of!Memory)!
Change!the!value!of!the!syndrome!to!the!memory!mode!ABAB!
```

6.6.3 Access to the Syndrome

The design of the TLMU should allow the possibility of neighbour elements accessing the memory elements. Synchronisation is then required to avoid several elements accessing the same memory at the same time.

Since the syndrome should also be accessible by other elements it may be a good idea that the syndrome is located in this interfacing area. Having access to the syndrome via TLMU would save logic since only one synchronisation element (TLMU) will be used.

6.7 Conclusions

- A resilient architecture was proposed including the hardware and the system software elements that can provide efficient performance, reliability and energy-smartness.
- The principles of designed followed and the structural properties of active, passive and interfacing zones are introduced. Active zone of hardware was also described with emphasis on recoverability after malfunctions and implementation of checking schemes. A processor with a reduced instruction set and a careful introduction of redundancy including checking schemes and re-execution at the instruction level can provide higher and efficient reliability.
- Reconfigurability of a real-time architecture at the system level was proposed and analysed in the context of each zone. With regards to the interfacing zone, a new element (T-logic), as a basic unit of reconfiguration, and its different configurations were proposed. We analysed and described how the flexibility of the T-elements has a positive effect in reliability and power-smart functioning of the system.
- System-level reconfigurability can be achieved using a new hardware element called *Syndrome* that can provide essential knowledge about hardware conditions. We showed the relation of this element with the active, passive and interfacing zones and how it can be used to implement GAFT. Functions of the *Syndrome* for reliability, performance and energy-smart functioning were described and explained.
- Taking into account that memory use has, by design, a high impact on system reliability and power consumption, passive zone reconfigurability was analysed and described in detail, including the control of configuration and the phases of hardware degradation.
- A Markov model of reliability for passive zone was developed, and analysis indicates the reliability gain of the different schemes permitted by the proposed reconfigurable architecture.
- System software support of testing and reconfiguration (dealing with system syndrome) was fully described. Shown that in combination of novel hardware architecture and system software all key properties of performance-, reliability- and energy-wise functioning can be improved.

Chapter 7
System Software Support

7.1 System Software Support of Hardware Checking

Consider a sequence of tests and programs T and P within a system as in Fig. 7.1. The initial test T is executed before a given task execution guaranteeing HW consistency, i.e. it guarantees that there is no fault at time t0 in the system.

However, if a permanent fault (e.g. a stuck bit) occurs immediately after the first test or during the program execution, it might be invisible for an arbitrary long time (latent period).

For periodic tasks, which are often used in control systems, we slightly adapt this scheme as shown in Fig. 7.2.

Nevertheless, what if a transient fault occurs during the execution of P? As mentioned in Sect. 3.5.2.1.1, at least three cases may take place:

- The effect of the fault lasts until P finishes, T detects the error and the recovery mechanisms are able to restore normal functioning on time (detected recoverable error or DRE);
- The effect of the fault lasts until P finishes, T detects the error but the recovery on time is not possible (detected unrecoverable error or DUE);
- The effect of the fault might not last until P finishes and therefore T would not be able to detect the fault, which in turn would allow the fault to remain undetected, perhaps causing data corruption (silent data corruption or SDC). The corruption could go unnoticed (benign error) or could result in a visible error.

Transient faults can be detected by re-executing the same program P with comparison C of the result and the result state space. Figure 7.3 illustrates this scenario. Note that for periodic tasks, the state of the program, which is used in the next computation as input data, must also be compared, as transient faults might affect data that is no longer used in the current computation but in the next.

Permanent faults however cannot be detected with the comparison scheme alone, as they might affect both executions of P. That is, the scenario in Fig. 7.2

V. Castano, I. Schagaev, *Resilient Computer System Design*,
DOI 10.1007/978-3-319-15069-7_7

Fig. 7.1 Ensuring HW integrity through program test execution

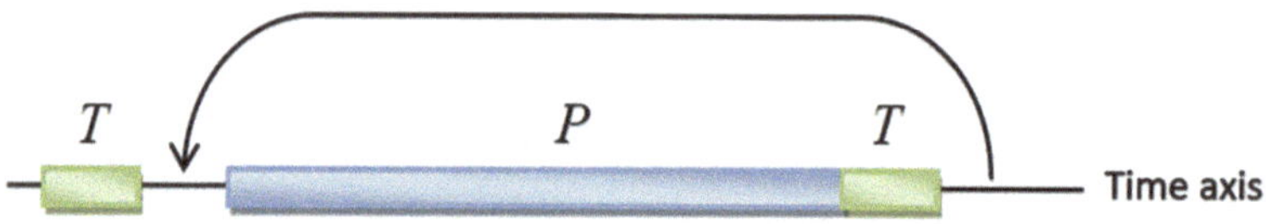

Fig. 7.2 Regular sequence of program execution with test of HW integrity to detect permanent faults

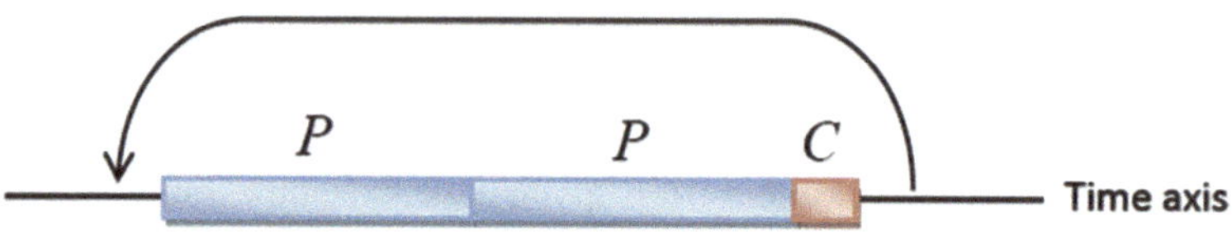

Fig. 7.3 Ensuring HW integrity through program test execution to detect transient faults

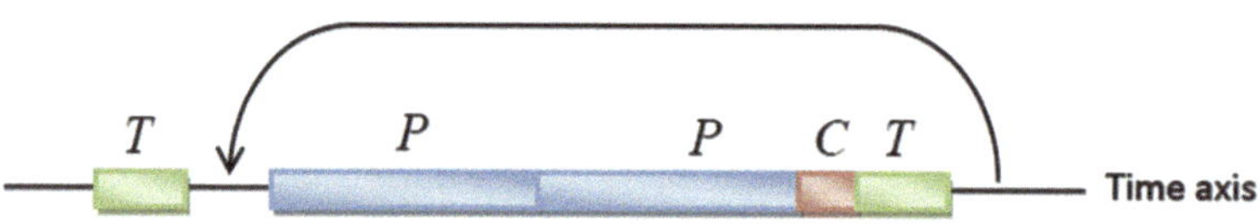

Fig. 7.4 Ensuring HW integrity through program test execution to detect transient and permanent faults

can detect only permanent faults, whereas the scenario in Fig. 7.3 can detect only transient faults.

The combined power to detect transient and permanent faults is illustrated in Fig. 7.4 [program and test sequence], where C is used to detect transient faults and T to detect permanent faults. Assuming that C triggers an error but T does not, it is clear that a transient fault has occurred. Another run of program P with comparison to the previous two runs can identify the run where the transient fault has occurred.

In the following analysis, we concentrate on the detection of permanent faults only and use only T in the analysis. The detection of transient faults can be considered as included in the following analysis if the double execution of P with following C as a whole is treated as task P in the following analysis.

A testing phase is required initially at boot up time to guarantee the correctness of the hardware and also a periodic test before and after the execution of a program. The applied tests might vary in depth (coverage), type of faults and the set of the tested hardware.

Every hardware component has typically at least one assigned test but might also have more than one that could differ on the implementation level. Software based tests need a processor and memory for the test execution even if a peripheral device is tested. In order to guarantee that faults in other hardware components that are not subject of the test itself do not have an influence on the outcome of the test, the order of the tests must follow the *principle of growing core*: if a test of a hardware component u_i has implicit dependencies on another hardware component u_j, the test of u_j must be executed first.

If the resources needed by a task are known in advance, it is sufficient to run after the execution only the testing procedures of the accessed hardware resources (selective testing), again by using the principle of growing core. This way, the system stays fully operational even in the case of present faults in some hardware components that are not in use. Spare components can be used for the relocation of programs that have been running on faulty hardware components.

Of course, it is also necessary independent of this scenario to periodically test the full hardware as otherwise faulty spare components are considered as fully operational and might be used again in a subsequent reconfiguration process.

A full hardware test also allows the system software to monitor the current full state of the hardware and take appropriate actions if necessary. If no spare components are available in the system, all programs depending on this component must be obviously terminated. If no essential program is affected by this component, the system can continue operating in a degraded mode.

For diagnostic and monitoring purposes the results of the tests should be available for the software or even external systems. We propose therefore to organise the test results of hardware-based test in so-called test syndromes before.

For every hardware component, for example the register file, AU, LU, internal bus or device controllers, the checking procedures present their result in the form of a syndrome to the software, indicating in binary form the state of the device. By grouping all syndromes together in one register, the software has a very effective way to check the integrity of the system. In case of a non-zero syndrome further analysis of the hardware conditions is required, especially when the malfunction duration is long.

Dependent on the used hardware checking scheme, it is not only possible to signal a fault to the runtime but also to provide more information to the runtime system to ease recovery. If for example the testing schemes discover stuck bits in memory, it is sufficient to recover programs that access the affected location and not all programs that are using this memory module.

Device drivers could for example provide their own testing schemes for their respective device. Especially for devices, one could think of having a combination of hardware and software based testing. I/O devices such as UARTs could effectively be tested by cross-connecting the input and output wires by very simple additional hardware logic and sending various bit patterns over this loopback connection.

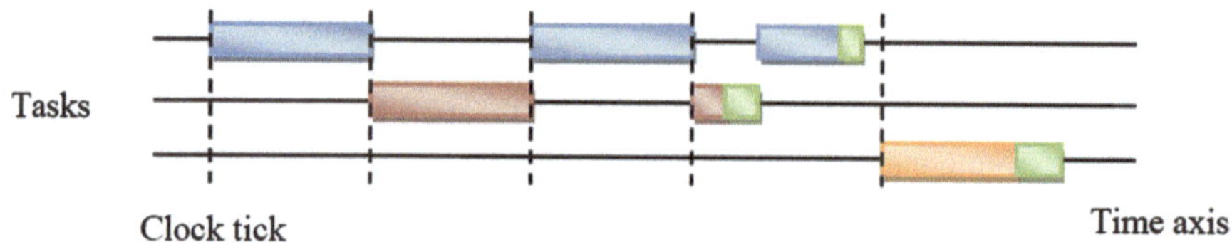

Fig. 7.5 Tasks and test combined

Timely task completion in real time systems is a key requirement, and therefore, the testing overheads should be reduced as much as possible when and where possible.

Figure 7.5 shows an example of three tasks with corresponding tests. Analysis of the checking process assumption in this case is based on a time slice scheduler, which distributes time slices to the running processes.

In this example, the processes run to completion and are called periodically by the scheduler. Three tasks are running, each with its own test (the green boxes) at the end of the task execution.

The test only checks the resources the respective process needs, which results in different test execution times. The task execution is only considered as successful if the test at the end of the task is successful.

If the test failed, the task is re-executed by using the same input data set as in the first try.

For diagnostic and monitoring purposes the test results should be available to the system software with supportive manifestation in syndrome that unit or to external systems. Therefore, we propose to organise the test results of hardware in test syndromes.

Difficulties arise if the task performs I/O on hardware devices or communicates with other tasks.

7.2 System Software Support for Hardware Reconfiguration

For every hardware component, e.g. register file, ALU, internal bus or device controller, the checking procedures present their result in the form of a syndrome to the software indicating, in binary form, the state of the device. By grouping all syndromes together in one register, the software has a very effective way to check the integrity of the system. In case of a non-zero syndrome further analysis of the hardware conditions is required, especially when the duration of the malfunction is long.

Depending on the checking scheme used, it is not only possible to signal a fault to the runtime but also to provide it with extra information to ease recovery. For instance, let us define a scenario where the testing schemes discover stuck bits in memory due to a Hard SEL. It would be sufficient to recover programs that access the affected location and not all programs that are using the affected memory module.

Device drivers could for example provide their own testing schemes for their respective device. Especially for devices, one could think of having a combination of hardware and software based testing. I/O devices such as UARTs could effectively be tested by cross connecting the input and output wires using very simple additional hardware logic and sending various bit patterns over this loopback connection.

In case of a detected hardware fault, the syndrome raises a hardware interrupt and SSW takes control of the reconfiguration process. The whole procedure is almost identical in the case of software-based schemes detecting the fault, with the difference that the interrupt that is raised is not hardware but software based. In the case of memory errors, if the current memory configuration does not use a redundant mode, software based checking is the only possible approach.

The general procedure of software support during reconfiguration is listed as follows:

- The hardware checking scheme triggers the syndrome interrupt.
- In order to distinguish the fault type, the syndrome interrupt handler then either initiates a *HW BuiltNIn SelfNTest* (*BIST*) procedure of the device or runs a SSW based self-test. In the case of SSW tests, writing different memory patterns to the faulty memory address can be used to derive the fault type. The syndrome bit indicating the fault has to be cleared after recovery. If after the test and recovery, the syndrome still shows the fault, the memory module is considered faulty. The affected memory address is still present in one of the processor registers and based on the IRQ return address, the correct register number can be derived by decoding the memory instruction that triggered the fault. In general, all software based testing procedures that test memory must not use the stack (no procedure calls, no data pushed on the stack) until the proper functioning of the used stack locations is ensured.
- In case of a transient fault, the event is logged and the program execution resumed. Logging the events is important as an accumulation of malfunctions in a module or a specified memory location could hint a potential permanent failure in the near future.
- In case of a permanent fault, the current memory configuration is extracted from the syndrome, and the next degradation state is calculated according to the application needs and predefined degradation tables.
- The new calculated memory configuration is written to the syndrome registers and the power of the faulty unit is removed. In some configuration transitions, the SSW has to adapt to the new situation and recover after excluding the faulty unit but before including the new one.
- SSW clears the fault in the syndrome and resumes processing by returning from the interrupt.

Some of the presented transitions in Sect. 6.5 need software intervention to fully recover from a permanent fault and to establish a new working software state. We show here a few situations where SW support is needed:

- *Adding/Replacing a module of an already populated bank*: if a memory module of a redundant memory mode (DMR or TMR) that is suspected of presenting faults is replaced by another module, memory content must be replicated to the new module before it is included in the configuration. A small routine following the algorithm described in Sect. 6.6.2 is sufficient to perform the copy without modifying the memory during the operation. Before performing that routine, SSW configures the syndrome to include a spare memory bank. After the copy operation, the spare module can be included in the working set. These actions must be performed for example when the system switches from Phase 2 to Phase 1 in Fig. 6.10.
- *Failed module in the runtime system area*: The area of memory that contains by convention all runtime system data structures is critical for system operation. When the recovery procedures are unable to overcome the situation, if the module has a replicated pair in redundant mode, the affected module is replaced by its replicated version via syndrome. However, if the module is not replicated, resetting the system either via a hardware watchdog or a software initiated power cycle is the option as a last resort. The BIST mechanism automatically identifies the failed module and configures another still working module to bank 1. The runtime system can then restart all critical applications.
- *Fault in a memory module that is not replicated*: this case is the most difficult to handle as the software must adapt to the smaller available memory space.

Instead of graceful degradation, software can also decide to "upgrade" the system in terms of redundancy, i.e. going from a mode with less redundancy to a mode with higher redundancy. The inclusion of a spare module corresponds to the first point in the list above; if an already used module is moved to another bank, software has to release all data structures residing on that module in case it is in non redundant use and repopulate it with data according to Point 1.

Intentional change of the operating mode to a less redundant mode is of course also possible, and needs no special software measures. As soon as the module is reconfigured to a free bank, the runtime system can start using it.

7.3 System Software Monitor of Hardware Condition

A hardware monitor, which is part of the runtime system, is responsible for keeping track of the hardware state. For every hardware component, which is managed by the syndrome, the hardware monitor tracks the state in more detail than the syndrome alone can provide.

It is also responsible for the execution of all software checking schemes and performs the actual hardware reconfiguration.

Thus, the hardware monitor must be accessible by the syndrome interrupt handler as well as the runtime system. This monitor should however not directly be accessible by applications; only drivers, which are part of the runtime system, can register checking procedures for their respective hardware component.

When the system is turned on, the BIST procedures embedded in the system are executed. It runs tests on all devices, using the principle of growing core, to ensure the integrity of all devices. If a failure is detected, the syndrome sets the appropriate fault bits.

The BIST is also responsible to initiate the system to a predefined working state, i.e. the most reliable mode with all working available resources.

When the BIST finishes and passes control to the runtime system, the runtime passes control to the hardware monitor which first mirrors the current state in software and then reconfigures the system according to the need of the program.

As the syndrome might trigger an interrupt right after boot up, the syndrome interrupt handler has to ensure that the stack pointer is valid and if not initialise it.

Every hardware component, which is managed by the syndrome, has shown with exactly one state of Fig. 7.6. This state diagram shows also all possible transitions between states, allowing the hardware monitor to reconfigure the system in a consistent way.

In fact, all of the above presented cases in the degradation scenarios where software intervention is required are clearly identifiable in Fig. 7.6.

Intervention is only required if the state transition goes from Stand-by to one of the active cases (marked in blue).

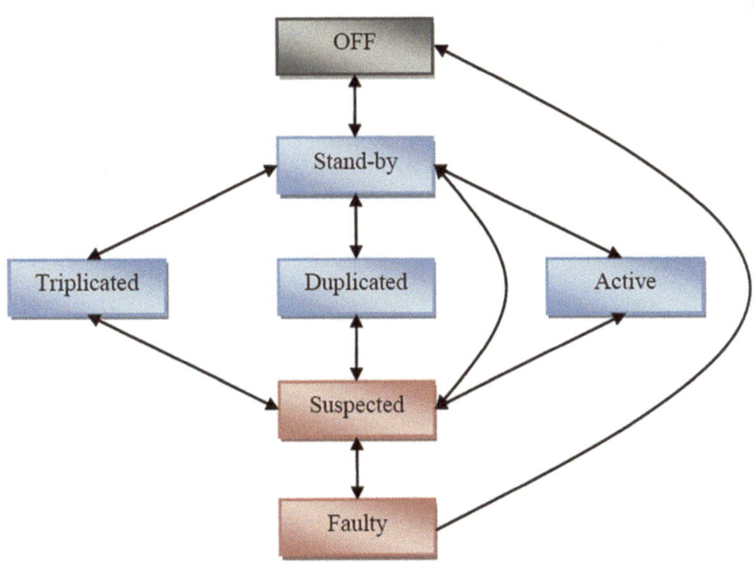

Fig. 7.6 Hardware state diagram

After boot up, all devices are either in state OFF or in one of the blue operation modes. As the BIST automatically configures the most reliable possible memory configuration, the initial states of all devices must be acquired by reading the syndrome. Given here is a short list of all possible states with a short description of them:

- ***OFF***: the device is currently not in use, powered o_ and isolated for fault containment
- ***Stand-by***: the device is powered on but not yet in use, i.e. in case of memory not yet assigned to a memory bank. In case of reconfiguration, all transitions go through this state.
- ***Active***: the device is in use in a non-redundant mode. In case of memory, the memory module is assigned to a bank in linear non-redundant mode.
- ***Duplicated***: the device acts in duplicated mode.
- ***Triplicated***: the device acts in triplicated mode.
- ***Suspected***: As soon as a fault in the hardware is detected, the state of the affected hardware component is set to "suspected" and the testing procedures are initiated to diagnose the fault. If a device is often in this state, this could be a hint that the device might fail in the near future. For reliability purposes it might therefore be sensible to replace the component with a spare one.
- ***Faulty***: dependent on the analysis outcome, the state is then set either to *Faulty* if a permanent fault has been diagnosed or back to the previous state if it has been only a transient fault. A device in the state Faulty is powered off.

The state transition diagram of Fig. 7.6 is not directly applicable to all devices. A memory bank for example should during the transition not go through stand-by to make sure that the stored data in the attached memory modules are not lost.

Periodical hardware checks should be performed on every single hardware component despite its state. This ensures that no hidden faults can stay undetected in the system and possibly spread. Thus, from any state exists a transition to the "suspected" state. Also, from Stand-by state—when device or element is not currently used they might be tested regularly.

It is even possible to revive faulty components, as for example environmental changes could allow a component to function correctly again.

7.4 Conclusion

- System software support of hardware fault tolerance is described along execution of various steps of generalised algorithm of fault tolerance.
- We show that in principle hardware conditions and states might be detected, evaluated and to some extent restored by system software without any support of hardware design.
- Various schemes of testing organisation for detection of hardware malfunctions, hardware permanent faults are presented, advantages (flexibility, ease of

upgrade) are mentioned as well as drawbacks (needs of time redundancy, much higher performance of hardware growth of complexity of run-time system).

- Shown that implementation of testing programs might be "imbedded" into run-time system making tests of hardware elements transparent for multitasking operations.
- A special hardware structure called Syndrome is introduced in the context of system software control of reconfiguration of hardware.
- System software run-time system structure, functions and features are discussed to support phases of GAFT implementation: testing, malfunction toleration and reconfiguration control.

Chapter 8
Implementation: Hardware Prototype, Comparisons, Simulation and Testing

8.1 Instruction Execution

Figure 8.1 shows the execution flow of the proposed microprocessor. As we mention in Sect. 6.2.1, the execution steps are similar to other RISC processors and since there is no pipelining all steps of one single instruction are executed within one memory cycle.

The fetching step loads an instruction from main memory storing it into the Instruction Register (IR). Since every instruction is 16-bit wide this step is required every second instruction. Decoding and Execution of the Instruction follows.

After execution of the first instruction, the second instruction from the Instruction Register (IR) can be executed without access to memory. This eases the speed gap between processor and memory and reduces their dependency. Finally, storage of the result takes place if the instruction execution has affected the content of registers, processor flags or any other processor state.

The processor has two internal fetching states (F1 and F2) that are required by the memory controller. Again, simplicity is not only applied to the processor but also to the memory controller. Both are designed to avoid possible stalling due to pending memory operations. This can be achieved by interleaving instruction fetches and memory operations.

The fetching step always loads two 16-bit instructions from memory into the internal IR. In sequential instruction execution, the compiler can schedule memory instructions in every second instruction slot where no instruction has to be loaded by the processor.

An example of this notion is illustrated Fig. 8.2, which is an extended version of the instruction execution flow in Fig. 8.1. In the F1 state, the processor fetches an instruction from memory; therefore, since the memory unit is busy during this cycle, it cannot execute at the same time an instruction that involves memory.

V. Castano, I. Schagaev, *Resilient Computer System Design*,
DOI 10.1007/978-3-319-15069-7_8

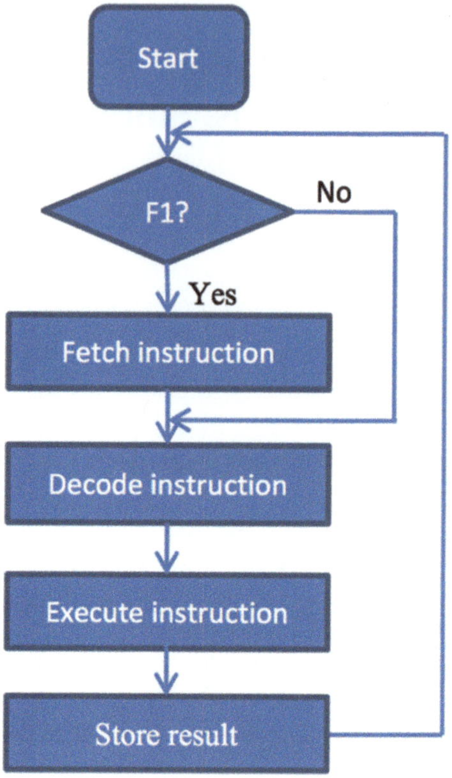

Fig. 8.1 Simple version of the prototype's Instruction execution

Only when the processor is in the fetching state F2, the processor is able to execute a memory instruction. It is the compiler's responsibility to schedule the instruction in the proper order.

We choose to simplify the hardware design at the expense of adding complexity to the compiler. By doing so, we reduce the amount of redundancy and therefore increase system reliability. If the instruction executed is a branch instruction the processor is switched automatically to the F1 state.

The reason for that is that the memory controller can only load 32-bit aligned addresses and therefore, the jump destination locations must also be 32-bit aligned. All these arguments and presented factors force the compiler to fill the memory gaps with NOP instructions.

8.2 Instruction Set

As mentioned in Sect. 6.2.1, in the ultra reduced instruction set employed, each one of the 16 instructions is encoded into 16 bits and only two of them are memory instructions (load/store).

Fig. 8.2 Instruction execution (extended version)

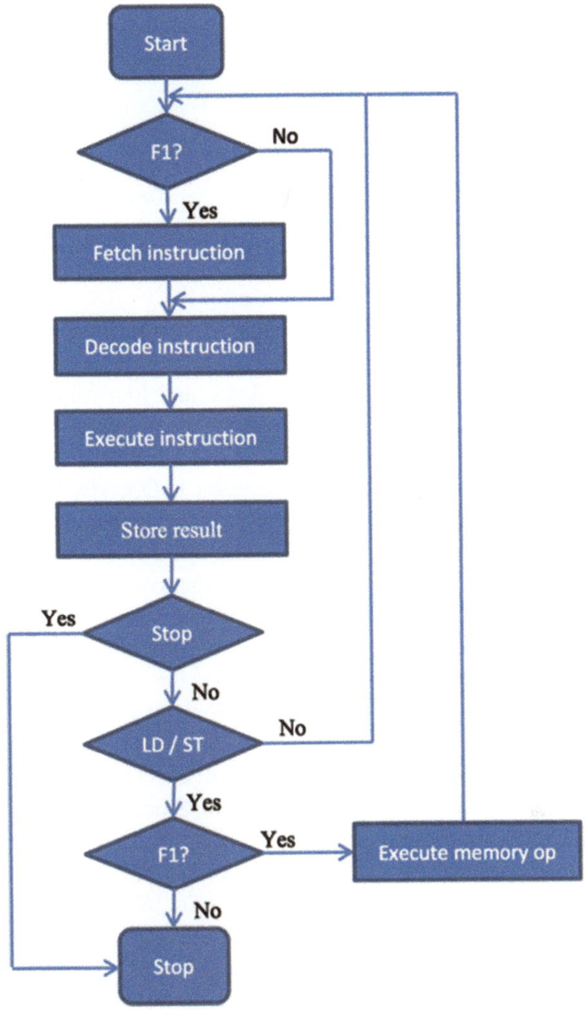

The instructions are designed as two operands instructions. They expect exactly two arbitrary registers as input, and save the result of the operation in the first register, thus overwriting one of the input values.

The compiler has to keep in mind that the content of the first register is overwritten. Again, we increase the simplicity of design at the expense of compiler's complexity. The impact of the size of the register file on overall performance of processor is a question of further research.

Figure 8.3 illustrates the instruction format divided in four different areas. Bits 15 and 14 (in red) indicate the format of the operation, which could be 8-, 16- or 32-bit. Bits 13–10 (in grey) contain one of the 16 different operation codes. Bits 9–5 (in blue) and 4–0 (in green) contain the first and second operand, which could be any of the 32 general purpose registers in the *RF*.

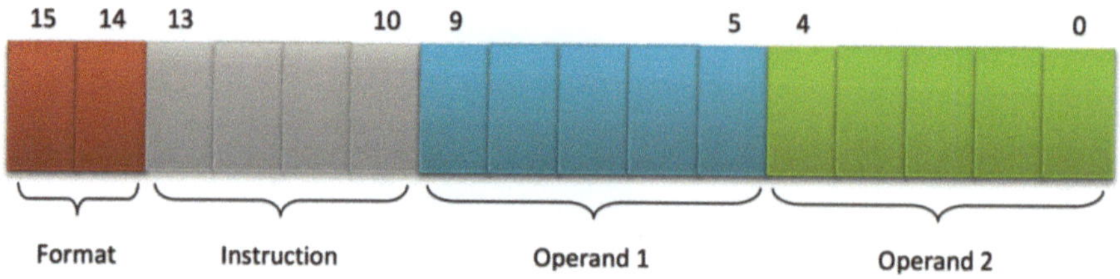

Fig. 8.3 Instruction format

The following Table 8.1 lists the current ISA with a short description of every instruction together with their assembler representation. The current architecture comprises seven control, five logic and four arithmetic instructions. Whilst most entries in Table 8.1 are self-explanatory, when we feel necessary we add some further explanation.

A special case of code operation is Opcode 0 which in combination with the format represents four different instructions: *STOP*, *NOP*, *TRACE* and *RETI*. *TRACE* is used for debug purposes and *RETI* is used to exit an interrupt handler.

The compiler needs to be aware that constants cannot be directly encoded in the instructions.

LDA is another special instruction that loads a constant to the specified register. The next aligned 32 bits aligned after the active program counter store the constant to be loaded. Placing two instructions into one 32-bit word increase code density and performance.

As explained earlier, the processor executes the first instruction on the left before executing the second one on the right. Since the program counter has the same value for both instructions it would be problematic to jump directly to a second instruction, which is not 32-bit aligned and has no unique address. Therefore, it is responsibility of the compiler to insert *NOPs* at the right places to prevent cases where instruction reordering fails to fill the gap.

The *CND* instruction performs an arithmetic comparison of the Registers *Ri* and *Rj* storing the result in *Rj*. The functioning is similar to other platforms: the comparison involves checking of three conditions saving these as flags in the first three bits of *Rj*:

- Bit 0: $R1 > R2$
- Bit 1: $R1 < R2$
- Bit 2: $R1 = R2$

Table 8.2 shows the relations of the comparison operations and their corresponding bitmasks. By applying an appropriate mask to the flags, the result for every possible comparison operation can be used as an argument in a conditional jump or saved as a Boolean value.

At the moment there is no support for unsigned operations. All arithmetic operations treat the values in the operands as signed values. All instructions accept the same register for both arguments with the exception of conditional jump instruction.

Table 8.1 Instruction set explained

Name	Format code	Op. code	Op1	Op2	Op2
NOP	01	0000	Ignored	Ignored	Execute no action except increasing the PC
STOP	00	0000	0	0	Stop instruction Execution
TRACE	00	0000	$Ri > 0$	$Rj >$	Output Ri to debugger. Operand1 or Operand2 must be >0
RETI	11	0000			Return from interrupt (Address in R31)
LD	11	0001	Ri	Rj	Load 32-bit memory at address Ri into Register Rj ($Rj := *Ri$)
LDA	00	0010	Ignored	Rj	Load the value from the next 32 bit words (rel. to PC) and store it in Rj ($Rj :=$ constant). Operand1 is ignored
ST	11	0011	Ri	Rj	Store content of register Ri to the memory at address Rj ($*Rj := Ri$)
MOV	XX	0100	Ri	Rj	Move content of register Ri to register Rj ($Rj := Ri$)
ADD	XX	0101	Ri	Rj	Arithmetically add the content of Ri to the content of Rj and store the result in Rj ($Rj := Rj + Ri$)
SUB	XX	0110	Ri	Rj	Arithmetically subtract the content of Ri to the content of Rj and store the result in Rj ($Rj := Rj - Ri$)
ASR	XX	0111	Ri	Rj	Shift the content of register Ri arithmetically one bit to the right and store the result in Rj
ASL	XX	1000	Ri	Rj	Shift the content of register Ri arithmetically one bit to the left and store the result in Rj
OR	XX	1001	Ri	Rj	Perform a bitwise logical OR of register Ri with register Rj and store the result in Rj
AND	XX	1010	Ri	Rj	Perform a bitwise logical AND of register Ri with register Rj and store the result in Rj
XOR	XX	1011	Ri	Rj	Perform a bitwise logical XOR of register Ri with register Rj and store the result in Rj
LSL	XX	1100	Ri	Rj	Shift the content of register Ri logically one bit to the right and store the result in Rj
LSR	XX	1101	Ri	Rj	Shift the content of register Ri logically one bit to the left and store the result in Rj
CND	XX	1110	Ri	Rj	Arithmetic comparison of Ri with Rj and store the result in Rj
CBR	XX	1111	Ri	Rj	Jump to address in Rj if Ri is non-zero and save PC in Ri

Table 8.2 Instruction set explained

Relation	Bit mask
$<$	010
$\leq$	110
$=$	100
$>=$	101
$>$	001
$\neq$	011

8.3 ERA Hardware Prototype

Figure 8.4 presents the current custom hardware prototype of the ERA device. The prototype is provided with two flash based ROM modules 128 Mb each (8 Mb × 16 bit) with replicated bootstrapping firmware and operating system software.

The proposed memory scheme may be regarded as a collection of 4 blocks, 16-bit wide, with identical size of 1 Mb each (16 × 64k). Using 16-bit memory modules instead of 32-bit memory modules increases reliability and reduces when necessary energy required for execution.

Reliability is increased by means of added working states and configurations.

Energy-wise operation is improving by means of this architecture ability to activate only modules required for application—by means of using a single 16-bit memory module, when necessary.

Table 8.3 presents a basic memory map with memory locations occupied by the ERA devices: ROM (2 banks) RAM (4 banks), USB, Ethernet, and UART interface. The 4 RAM modules are 1 Megabit each (64k × 16 bit). The flash based ROM modules are 128 Mbit each (8 Mb × 16 bit).

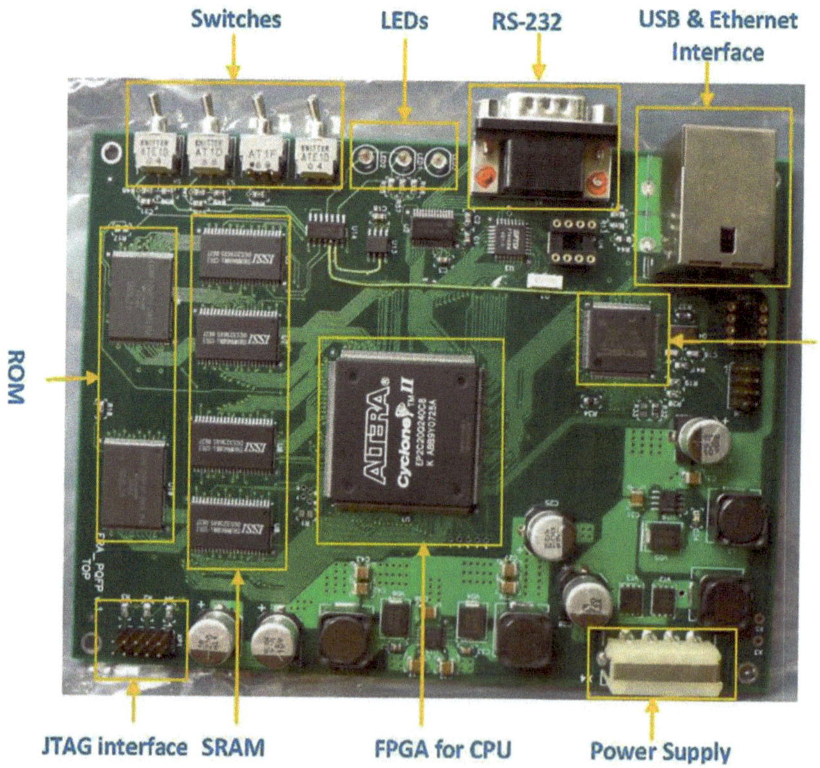

Fig. 8.4 ERA prototype board

Table 8.3 ERA memory map

Memory range	Device details	Device
0700FFFFH–07000000H	ISSI-IS61WV6416BLL	SRAM logic module 2 (U7, U8)
0600FFFFH–06000000H	ISSI-IS61WV6416BLL	SRAM logic module 1 (U5, U6)
057FFFFFH–05000000H	Sharp-LH28F128BFHT	ROM logic module 1 (U9, U10)
04000000H	FTDI-FT245BM	USB
0300FFFFH–03000000H	SMSC-LAN91c11i	Ethernet
02000002H–02000000H	RS232	UART Interface
01000000H	Normal	LEDs (1–2)
00000000H	KNITTER	Switch (3–4)

The SRAM modules representing the units 8, 7, 5 and 6 are located in the highest part of the memory, followed by the ROM modules (units 9 and 10). The USB, Ethernet, Serial Ports, LED's and switches of the ERA board are mapped in the lower part of memory.

8.4 Architectural Comparison

Nowadays the embedded processor market is dominated by the ARM architectures with their RISC processors. Other relevant hardware architectures are the LEON designs that include FT versions of their SPARC processor.

Although 64-bit versions are available for the ×86 and ARM architectures, in order to keep consistency, we have chosen to make a comparison of 32-bit version processors including Intel ×86's architectures.

Table 8.4 provides an overview of these hardware architectures and their driver features. The table is based on data gathered from (Gaisler 2002; Heise 2009; Hennessy and Patterson 2006; Sparc International, Inc. 1998).

SPARC and ARM processors are based on a simple Reduced Instruction Set Architecture and therefore more similar to the ERRIC processor, whilst the ×86 is based on a Complex Instruction Set with a much larger number of instructions. The table clearly shows the simplicity of ERRIC's ISA with its 16 instructions, which is by large margin smaller than the RISC and CISC architectures. Simple and less powerful instructions come with the cost of longer code compared to the other platforms.

The ×86 is a *registerNmemory* architecture that allows using memory locations directly in instructions. Conversely, ERRIC, ARM and SPARC as *load and store* architectures must first load the argument into a register. The enormous number of instructions of ×86 has lead to the situation where the instruction decoder of an Intel Atom CPU occupies more chip's real state than the complete ARM Cortex-A5 (Heise 2009).

As an example let us examine the load from memory.

Table 8.4 ERA comparison of hardware architectures

	ERRIC	x86	SPARC v8	ARM7TDMI (ARMv5-TE)	ARM7TDMI thumb
ISA type	MISC	CISC	RISC	RISC	RISC
Integer registers	32 × 32 bits	8 × 32 bits	31 × 32 bits	15 × 32 bits	8 × 32 bits + SR, LR
Floating point registers	0	Optional 8 × 32 bits or 8 × 64 bit (8 × 80 bits internal)	32 × 32 bits or 16 × 64 bits or 8 × 128 bits	Optional 32 × 32 bits or 16 × 64 bits	Optional 32 × 32 bits or 16 × 64 bits
Vector registers	0	Optional 8 × 64 bits or 8 × 128 bits	0	Optional 32 × 32 bits or 16 × 64 bits	0
Address space	32 bits flat	32 bits, flat or segmented	32 bits flat	32 bits flat	32 bits flat
Instruction size (bytes)	2	1–15	4	4	2
Multi-processor capable	No	Yes	Yes	Yes	No
Processor modes	1	3	2	7	7
Data aligned	Yes	No	Yes	Yes	Yes
MMU	Yes	Yes	Optional	Optional	Optional
Memory addressing modes	1	7	2	6	6
Memory addressing sizes	32-bit	8, 16, 32	8, 16, 32, 64	8, 16, 32	8, 16, 32
ISA size	16	332:138 Integer and logic, 92 floating point	72	53	37
I/O	Memory mapped	Instructions, memory mapped	Memory mapped	Memory mapped	Memory mapped
Pipeline length	No pipeline	Atom: 16 i7: 14 Pentium4: 20–31	Leon3: 7 SPARC64V:15 Ultra-SPARC T2: 8	3	3
Specialities	Very simple ISA, built-in FT	Big ISA and memory operands	Register window, delayed control transfer	Conditional instruction execution	32-bit ARM instructions partly required

Prior to the memory access, e.g. in the case array accesses, the absolute memory address must be explicitly calculated and stored in a register. In the SPARC architecture the offset to the base address can first be stored in a distinct register and then added on the fly in the load instruction itself. ARM processors even permit to encode an offset to a base address given in a register directly in the instruction itself.

The instruction set of ARM Thumb is a subset of the standard 32-bit ARM ISA. The Thumb's version targets resource constraint environments where only a 16-bit data bus is available. The address space of the Thumb's is still 32-bit and all registers are 32-bit wide.

Nonetheless, while R0–R7 are directly accessible, R8–R16 are hidden. ERRIC's ISA, although even more constraint than the Thumb's, is intended to be used as a full instruction set, generic enough to encode all language features.

Compared to ARM's, the ERRIC is much simpler and counts with less than half the number of instructions. However, with ERRIC more elaborate instructions need to be emulated, e.g. relative memory accesses or procedure calls.

The latter in particular is very efficiently implemented on the ARM by the "load multiple" and "store multiple" instructions, which allows to efficiently putting all procedure arguments on the stack. In contrast, ERRIC has to individually store all arguments on the stack.

Table 8.5 illustrates the different data addressing modes supported by the compared architectures. In some cases, such as SPARC, the architecture does not directly provide an absolute addressing mode. In order to emulate absolute addressing, the SPARC microprocessor uses a register + register mode with a nullified second register. It even provides a register, which is always nullified, so that the absolute addressing emulation does not incur any performance penalty.

CISC addressing modes are more powerful than RISC ones. Figure 8.5 summarises the $\times 86$ addressing modes. The offset part of a memory address, can be either a static displacement or through an address computation made up of one or more elements. The resulting offset is called effective address and can constitute either positive or negative values except for the scaling factor.

CISC addressing modes are more powerful than RISC ones. Figure 8.5 summarises the $\times 86$ addressing modes. The offset part of a memory address, can be either a static displacement or through an address computation made up of one or more elements.

Table 8.5 Supported addressing modes

Data addressing mode	ERRIC	x86	SPARC v8	ARM	ERM Thumb
Register	X	X	(X)	X	X
Register + offset (displacement or based)	–	X	X	X	X
Register + register (indexed)	–	X	X	X	X
Register + scaled register (scaled)	–	X	–	X	–
Register + offset and update register	–	–	–	X	–
Register + register and update register	–	–	–	X	–

$$offset = \begin{Bmatrix} CS: \\ DS: \\ SS: \\ ES: \\ FS: \\ GS: \end{Bmatrix} \begin{bmatrix} EAX \\ EBX \\ ECX \\ EDX \\ ESP \\ EBP \\ ESI \\ EDI \end{bmatrix} + \left[\begin{pmatrix} EAX \\ EBX \\ ECX \\ EDX \\ EBP \\ ESI \\ EDI \end{pmatrix} * \begin{pmatrix} 1 \\ 2 \\ 4 \\ 8 \end{pmatrix} \right] + \begin{bmatrix} None \\ 8-bit \\ 16-bit \\ 32-bit \end{bmatrix}$$

Fig. 8.5 Addressing modes of the ×86 architecture

Table 8.6 Offset sizes encoded in instructions

Offset size encoded in instructions (in bits)	ERRIC	x86	SPARC v8	ARM	ERM thumb
Unconditional jump call	0	8–32 Signed, relative or absolute, direct or indirect	30	24	11
Conditional branch	0	8–32 Signed	19	24	8

The resulting offset is called effective address and can constitute either positive or negative values except for the scaling factor. ERRIC provides only absolute memory addressing. Since the addresses must be explicitly computed before the data can be loaded, ERRIC's code requires more instructions. Table 8.6 compares the offsets sizes that are directly encoded in the instruction.

Unless mentioned differently, the offset is always relative to the current instruction pointer. ERRIC does not allow encoding the offset in an instruction. Instead, the offset is always given as an absolute address in a register. Therefore, ERRIC's requires an additional "load constant" instruction, which involves another extra 8 bytes (*Load + NOP + 4 Byte constant*).

With regard to safety of code, absolute jumps are desirable to relative jumps. The reason for that is that calculated jumps (relative) are more prone to faults than absolute jumps.

In the following page Table 8.7 presents an overview of the instructions that are required in the compared architectures to be able perform basic operations (such as load and stores). If an instruction is not available in the ISA, a sequence of instructions is given. In Table 8.7, a "–" sign means that in order to simulate the specific functionality more than a short instruction sequence is required. All floating-point instructions are omitted, as the ERRIC architecture does not include a floating-point unit, and all of them are emulated in software.

Table 8.7 Comparison of selected instructions

Instructions	ERRIC	x86	SPARC v8	ARM7TDMI (ARMv5-TE)	ARM7TDMI thumb
Load word	LD	MOV	LD	LDR	LDR
Load byte signed	-	MOVSX	LDSB	LDRSB	LDRSB
Load byte unsigned	LD, LDA, AND-[a]	MOV	LDUB	LDRB	LDRB
Store word	ST	MOV	ST	STR	STR
Store byte	-[b]	MOV	STB	STRB	STRB
ADD	ADD	ADD	ADD	ADD	ADD
ADD (trap if overflow)	-	ADD, INTO	ADDcc, TVS	ADDS, SWIVS	ADD, BVC +4, SWI
Sub	SUB	SUB	SUB	SUB	SUB
Sub (trap if overflow)	-	SUB, INTO	SUBcc, TVS	SUBS, SWIBS	SUB, BVC +4, SWI
Multiply	-	MUL, IMUL	MULX	MUL	MUL
Divide	-	DIV, IDIV	DIVX	-	-
AND	AND	AND	AND	AND	AND
OR	OR	OR	OR	ORR	ORR
XOR	XOR	XOR	XOR	EOR	EOR
NOT	-	NOT	-	-	-
Shift local left	LSL[c]	SHL	SLL	LSL	LSL
Shift local right	LSR[c]	SHR	SRL	LSR	LSR
Shift arithmetic right	ASR[c]	SAR	SRA	-	-
Compare	CND	CMP	SUBcc r0	CMP	CMP
Conditional Branch	CBR	CALL	CALL	BL	BL
Call	CBR	CALL	CALL	BL	BL
Trap	CBR	INT n	TIcc, SIR	SWI	SWI
Return from Interrupt	RETI	IRET	DONE, RETRY, RETURN	MOVS pc, r14	-[d]
NOP	NOP	NOP	SETHI r0, 0	MOV r0, r0	MOV r0, r0

[a]A sequence *LD*, *LDA*, *AND* must be used if the 8-bit data is aligned. Otherwise an LD and a specific number of shift operations must be used.
[b]In order to store an 8-bit value the destination address must be loaded, the appropriate bits must be cleared using a bit mask, the argument must be shifted and the written back to memory. It seems clear that omitting the use of 8-bit values would be more efficient.
[c]Only one bit.
[d]Since Interrupts are always handled in 32-Bit mode, and therefore, a pure 16-bit Thumb CPU would not support them.

8.5 ERA Testing and Debugging

Detailed explanation of testing and debugging steps are presented in Appendix A.

8.6 ERA's Assembler

The system software for ERA assembler-level programming consists of the following components:

- An **Assembler** that performs the compilation of source programs written in ERA assembler into executable or object code.
  - Program name: *Assembler.exe*
  - Call form: *assembler* + [+*options*+] + *source.era*
  - Result: *source.obj* + or *source.code*
- A **Linker** that takes several files with object code as input and produces the single result object or executable file depending on the existence of external references in the result.
  - Program name: *Linker.exe*
  - Call form: *linker* + [*options*] + [*entry_point*] + *source1.obj*...*sourceN.obj*
  - Result: *source1.code*
- A **Runner** that takes an ERA executable file as input, loads it into the memory (using the interface of the **Model** component) and executes it in the simulation mode.
  - Program name: *Runner.exe*
  - Call form: *runner* + [+*options*+] + *source.code*
  - Result: an output on the console or printing device
- A **Preparator**, an extra component which supports transition from the model to the real ERA board. The Preparator takes the ERA executable file as input and produces the pure binary file which is completely ready to load to the real memory.
  - Program name: *Preparator.exe*
  - Call form: *preparator* + *source.code*
  - Result: *source.bin*

The overall configuration of the assembler (except the Preparator) is shown on the picture below (Fig. 8.6).

As an example, below is source code of program implementing the simple in-place sorting algorithm. The program is written in the ERA assembler.

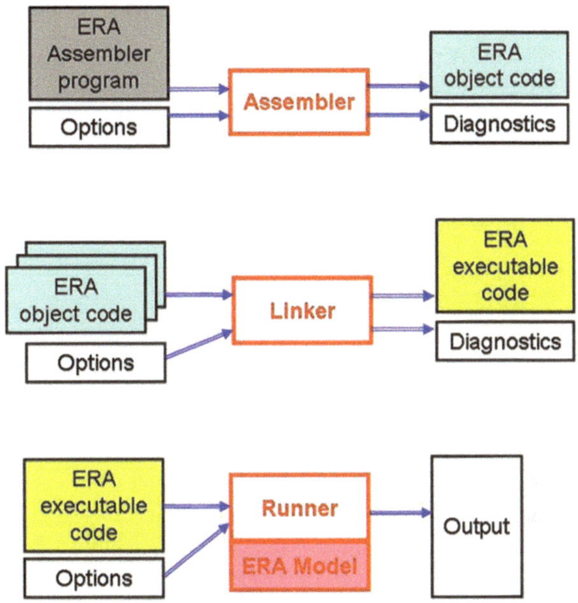

Fig. 8.6 Flow of ERA standard testing (*top*) and flow of testing with support of disassembler

The assembler's syntax for every EAR instruction was described before. The sorting algorithm itself is specified in Pascal-like language as follows:

```
procedure Sort ( a : array of int; size : int )
begin
    for i := 1 to size
       for j := i to 0 by -1
          if a[j-1] < a[j]
          then
             w := a[j-1];
             a[j-1] := a[j];
             a[j] := w;
          end
       end
    end
end Sort;
```

Additional remarks about ERA assembler syntax are:

- For debugging purposes two pseudo-instructions have been added to the ERA assembler: DATA and TRACE. DATA instruction just denotes literal data which will go directly to the object code. TRACE instruction causes registers specified in the instructions to be output to the console or to a printing device.
- Labels are specified as identifiers enclosed in angle brackets. The value of a label is the address of the instruction or data immediately following the label.

- Comments have the form // sequence of any characters until end of line

The ERA program implementing the sorting algorithm looks like as follows:

```
// i: R1;  j: R2;  intermediate values: R3, R4, R5

      R1 := 1;                  // i := 1
<LoopOuter>
      R3 := Size;
      R3 := *R3;                // R3: Size
      R4 := 1;
      R3  - = R4;
      R3 ?= R1;                 // Compare Size-1 and i
      R4 &= R3;                 // Extract > sign using R4=1 as mask for >
      R3 := Out Outer;
      if R4 goto R3;            // if i>Size goto OutOuter

      // Organize inner loop
      R2 := R1; NOP;            // j := i
<LoopInner>
      R3 := 0;                  // w := 0
      R3 ?= R2;                 // Compare j with 0
      R4 := 4;                  // Mask for equality
      R3 &= R4;                 // Extract equality sign
      R4 := Out Inner;
      if R3 goto R4;            // if j=0 exit the inner loop

      // Otherwise, compare two array elements
      // to decide if we need to exchange them.
      // R10: address of j-th element
      // R11: a[j]
      // R12: a[j-1]
      R10 := Array;
      R10 += R2;                // array base address+j
      R11 := *R10;              // R11 := a[j]
      R12 := R10;
      R13 := 1;
      R12  - = R13;             // a+j-1
      R12 := *R12;              // R12 := a[j-1]

      R3  := R12;               // w := a[j-1]
      R3  ?= R11;               // Compare a[j-1] and a[j]
      R4  := 5;                 // Mask for >=
```

```
        R4  &= R3;              // Extract > and = signs
        R3  := OutExchange;
        if R4 goto R3;          // if a[j-1] >= a[j] do not perform exchange

        // Otherwise, perform exchange
        R3 := R10;
        R4 := 1;
        R3 -= R4;               // R5: address of (j-1)th element
      *R3 := R11;               // a[j-1] := a[j]
      *R10:= R12;               // a[j] := a[j-1]
<OutExchange>
        // Decreasing j (inner loop)
        R3 := 1;
        R2 -= R3;               // j := j-1
        R4 := LoopInner;
        if R3 goto R4;          // goto LoopInner
<OutInner>
        // Increasing i (outer loop)
        R3 := 1;
        R1 += R3;               // i := i+1
        R4  := LoopOuter;
        if R3 goto R4;          // goto LoopOuter
        NOP;

<OutOuter>
        R15 := Size;
        R15 := *R15;
        R16 := Array;
        TRACE R15,R16;
        STOP; NOP;
<Size>
        DATA  20
<Array>
        DATA 537
        DATA 242
        DATA 114
        DATA 436
        DATA 337
        DATA 296
        DATA 285
        DATA 655
        DATA 639
        DATA 436
        DATA 912
        DATA 520
        DATA 624
        DATA 551
        DATA 600
        DATA 741
        DATA 612
        DATA 943
        DATA 871
        DATA 735
```

Table 8.8 Example of code transformed into assembly code by the assembler

Example 1. Global data and code				
Source code	**Memory structure**	**Code**	**Assembler. Code**	**Comments**

`char ch;`
`short int i;`
`int j;`
	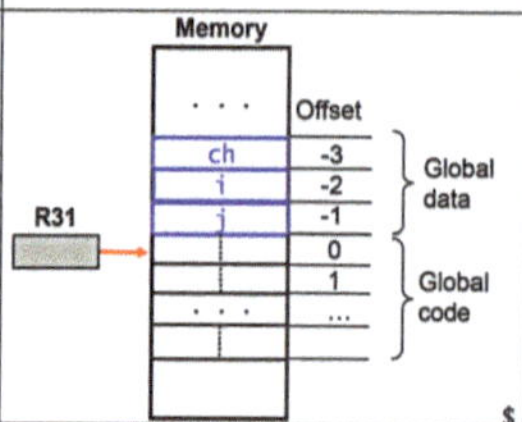
	
		$	$	Programming convention 1:
				R31 register always keeps the base address of the global data (with negative offsets) and the program code (with non-negative offsets).
				Initially R31 is set by the progam loader.

`ch := '0';$`	$	LDA R1 / NOP / '0'	R1 := '0';	Get the value of '0' into R1
		LDA R2 / ADD R31,R2 / -3	R2:=-6; R2+=R31;	Get the address of ch into R2 (as R31+offset)
		ST R1,R2	*R2:=R1;	Store the value from R1 to ch (pointed to by R2)
`i := 10;$`	$	LDA R1 / NOP / 10	R1:=10;	Get the value of 10 into R1
		LDA R2 / ADD R31,R2 / -2	R2:=-4; R2+=R31;	Get the address of i to into R2 (as R31+offset)
		ST R1,R2	*R2:=R1;	Store the value from R1 to i (pointed to by R2)
`j := i;$`	$	LDA R1 / ADD R31,R1 / -1	R1:=-2;	Get the offset of j into R1; get the address of j into R1 (as R31+offset)
		LDA R2 / ADD R31,R2 / -2	R2:=-4; R2+=R31;	Get the offset of i into R1; get the address of I into R2 (as R31+offset)
		LD R2,R2 / ST R2,R1	R2:=*R2; *R1:=R2;	Get the value pointed to by R2 (i.e., i) to R2. Store the value from R2 to j (pointed to by R1)

Another example of a simpler program is illustrated in Table 8.8. The table shows the location of data variables and code within the memory structure together with an explanation of the specific line of code and their effect. The *R31+* register, set by the program loader, always keeps the base address of the global data and the program code. The register uses negative offsets for the data and non-negative offsets for the code and local data.

8.7 ERA's Simulator: Dissimera

Reading binary code is a painful experience. In order to test and troubleshoot any error of design, bug or incompatibility between the assembler and the VHDL code, a new tool has been develop: a Disassembler and a Simulator in a combined tool that will ease this process. In addition it will allow the simulation of the state of the processor at any given time. Dissimera's main goal is to simulate the basic core features of ERA in a reliable and accurate manner.

The fundamental characteristics of this tool are:

- Disassembling of instructions: Binary-to-ASM and Binary-to-PSEUDOCODE that will complement the assembler
- Ability to discern data from instructions
- Simulation of the ERA architecture including: Program Counter (PC), Instruction Registry (IR), Register FILE (RF) and memory contents
- Step-by-Step Execution

- Breakpoints
- Overflow warning
- Logging
- Ability to compare results of simulation execution with the results of Altera execution.

8.7.1 Architecture and Description

The main two functions of Dissimera, Disassembling and Simulation, are embedded into a single software product. The architecture of such software is based on three main modules: the Interface module, the parsing module and the simulation module. The programming language used to implement those is ANSI C and currently targets 80 × 86 machines.

The Dissimera disassembler reads the output of the assembler introduced previously. Dissimera uses such output, in binary code, as an input, to start the simulation process.

The ***Interface module*** (***IM***) is based on an *NCurses* API with MIT licence to implement the interface under a Windows Console. The IM is built independently from the Dissimera's engine (composed by the parsing and simulation modules) with the intention of improving scalability. It is integrated in a way that escalating to a newer interface or migration to a different operating system will not be problematic.

Figure 8.7 illustrates the different elements of the current design of Dissimera's interface.

The current version is based on a single ERRIC's processor with a 32-bit mode. The *Paddr1* and *MAIN1* elements contain the addresses and binary contents of the 32-bit memory unit that contains the running code. *Paddr2* and *MEM* also contain a copy of the addresses and binary content but allow the user to browse the memory contents during and after execution and check the program results. *IDHEX3* shows the Instructions in assembler or the data in hexadecimal numbering. *ADDRF* and *RF* contain the name and value of the 32 registers that compose the register file. *PC* is the program counter or instruction pointer. The Instruction Register or *IR* stores the 32-bit value with the decoded instructions that are about to be executed. The status of execution and a log with extra details is shown in the STATUS element.

The ***parser module*** (***PM***) involves three different processes. The first process is the *lexical analysis* by which the input binary code is fragmented into meaningful symbols (tokens) in the context of ERA's pseudo-code language.

The next process is the *syntactic analysis* of these tokens that define allowable expressions according to the rules of a formal grammar, based on the ISA format introduced in before.

Finally, a *semantic analysis* works out the implications of the validated tokens and takes appropriate action.

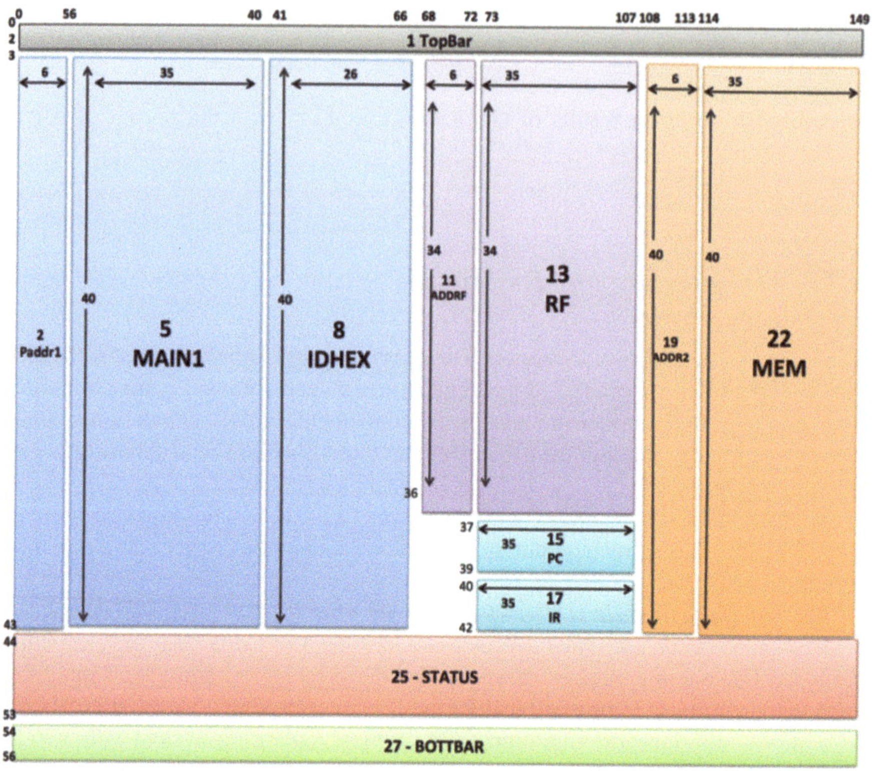

Fig. 8.7 Design of the interface of current version of simulator

These three processes need extra attention. How can we determine the type of a specific value? i.e. How can Dissimera be certain that a 16-bit binary value is either code or data?

Figure 8.8 depicts what we already mentioned, the program loader sets R31 with the base address of the global data (negative offsets) and program code (positive offsets).

However, in the case of local data, by just examining a single binary value it is not possible to determine its type, e.g. according to ERA's ISA rules, a *11000100111000013* value can be interpreted:

- as a *C4E1h* data value, or
- as an *LD3 R73 R13* instruction that loads the 32-bit value of the memory address contained in R7 into the register R1 (i.e. R1:=*R7).

The type of a value is determined by its context. Dissimera uses the execution context to determine that, by performing two top-down runs before the assembly code is presented on the screen. During the first run the code is fragmented and the starting point of execution is determined.

The determination of type takes place in the second run. Dissimera proceeds to silently execute instruction-by-instruction, marking the values as code tokens and

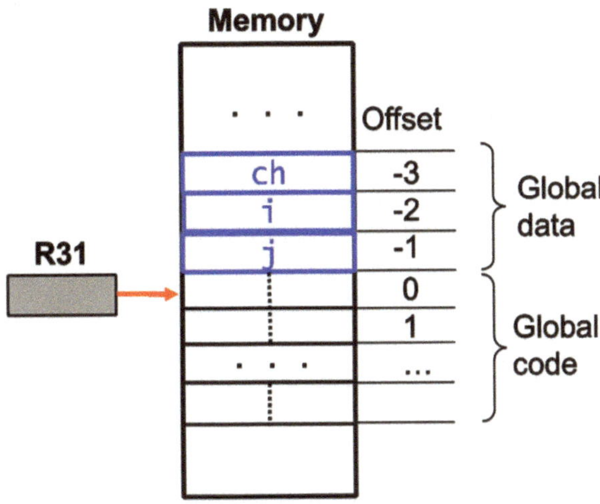

Fig. 8.8 Memory allocation of a program in ERA

decoding them consecutively performing the appropriate jump instructions and following the Program Counter.

Once all the code lines are executed the rest of the tokens are marked as local data tokens. This second run also includes error detection mechanisms for bad syntax and buffer overflow.

Note that both runs are transparent in terms of user interface. Once the parsing module has finished these processes, the UI contains now the results of disassembling and the simulation process can start.

The ***simulation module*** (*SM*) is in charge of program execution. This stage benefits from the two previous runs using the output of the parsing stage as an input. Hence, the SM is able to differentiate from code tokens and local and global data tokens.

Below, Fig. 8.9 presents a screenshot of the current version of Dissimera and the IM, PM and SM modules.

Initially, the execution starts at the address set by program counter (loaded with the content of R31). The *PC* holds the memory address of the next instruction to be executed and is incremented just after fetching the 32 bits from memory containing two instructions.

After the processor fetches the memory location stored in the PC, the instruction is loaded in the Instruction Register (*IR*). The instructions are fetched sequentially from memory unless a *CBR* instruction changes the sequence placing a new value in it.

Dissimera offers several run modes:

- *Normal Mode*: It is the standard mode. The simulator continuously executes instructions of a program until a STOP instruction is found.
- *Debugging Mode*: which includes the ability to place breakpoints within the code and the ability to perform step-by-step execution. In addition, this mode allows for step-back execution. Simulation can return to a previous state. All this features benefit the debugging of the system.

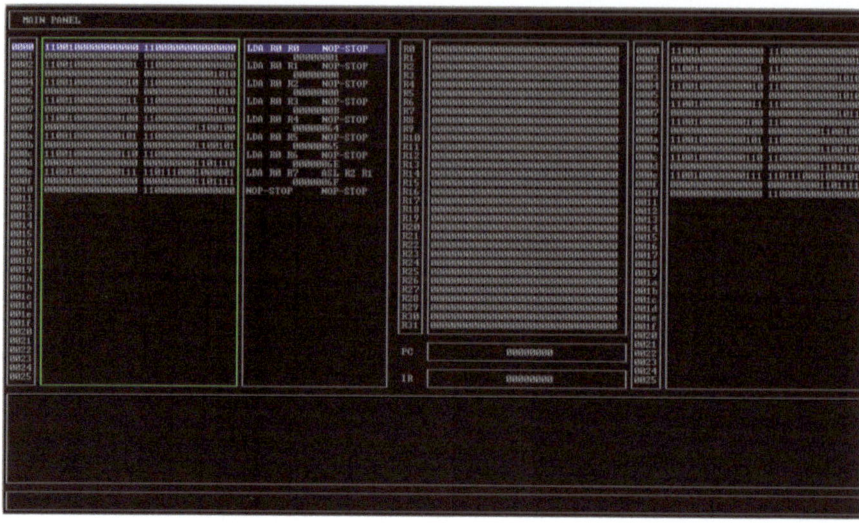

Fig. 8.9 Memory allocation of a program in ERA

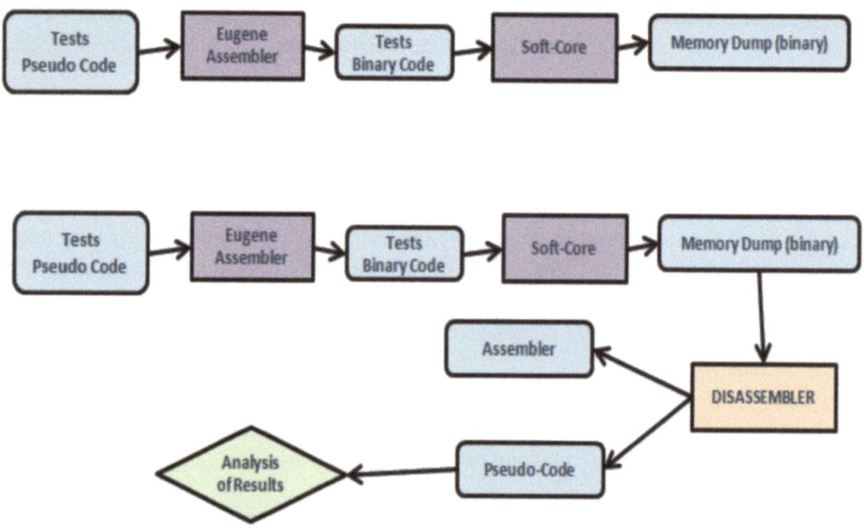

Fig. 8.10 Flow of standard ERA testing (*top*) and testing with Dissimera

Dissimera can be used as a tool for analysis and debugging of ERA programs, and more importantly, as a tool for testing and debugging of the hardware architecture. Figure 8.10 depicts two different testing methodologies for ERA. Initially, several test programs are developed using the pseudo-code language introduced before. The *assembler* is then used to obtain the tests programs in binary code compatible with ERA. Note that box "Eugene Assembler" defines an involvement in the debugging process system software written by Dr Eugene Zoueff, ETHZ, our collaborator in various projects and programs.

These test programs are introduced into ERRIC's soft-core processor through the JTAG Interface of ERA's board. The tests programs are executed with the Quartus II Software and upon execution a memory dump with the results of execution is produced.

After the Soft Processor simulation, the same initial test programs generated by the assembler are introduced into Dissimera together with the memory dump of execution results produced by Quartus II.

Here, the PM of Dissimera performs the disassembling of the binary code. After full execution of the ERA program by the SM, using the Dissimera's IM, the memory dump of the MEM element can be compared and analysed with the previous results of Quartus II.

In case of mismatch, debugging via step-by-step execution of both simulators can help in the detection and location of design and implementation errors.

In the following page, Figs. 8.11 and 8.12 show the caller graph of Dissimera's main function.

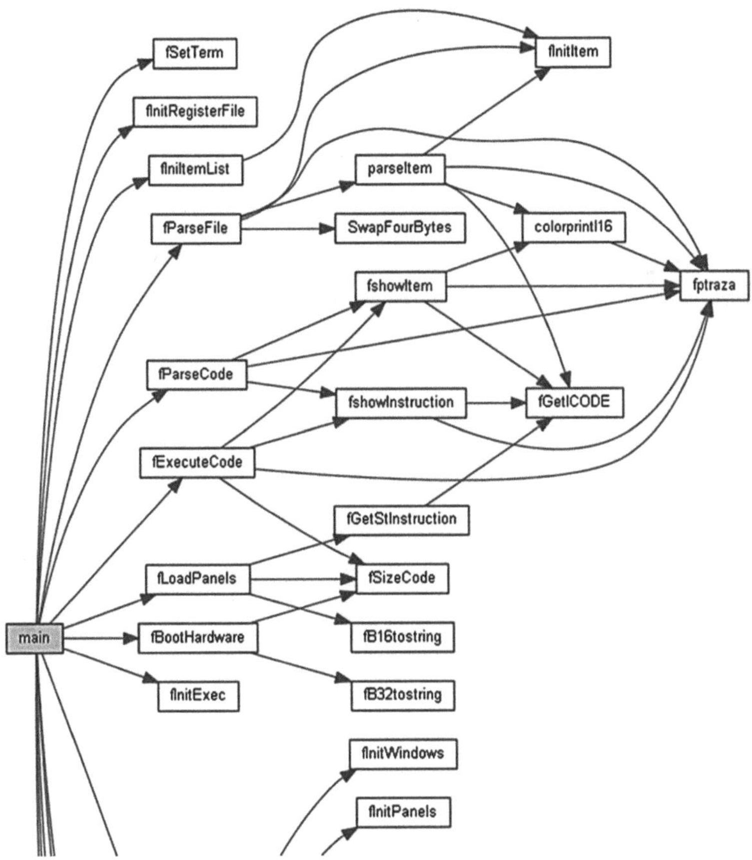

Fig. 8.11 Flow of standard ERA testing (*top*) and testing with Dissimera

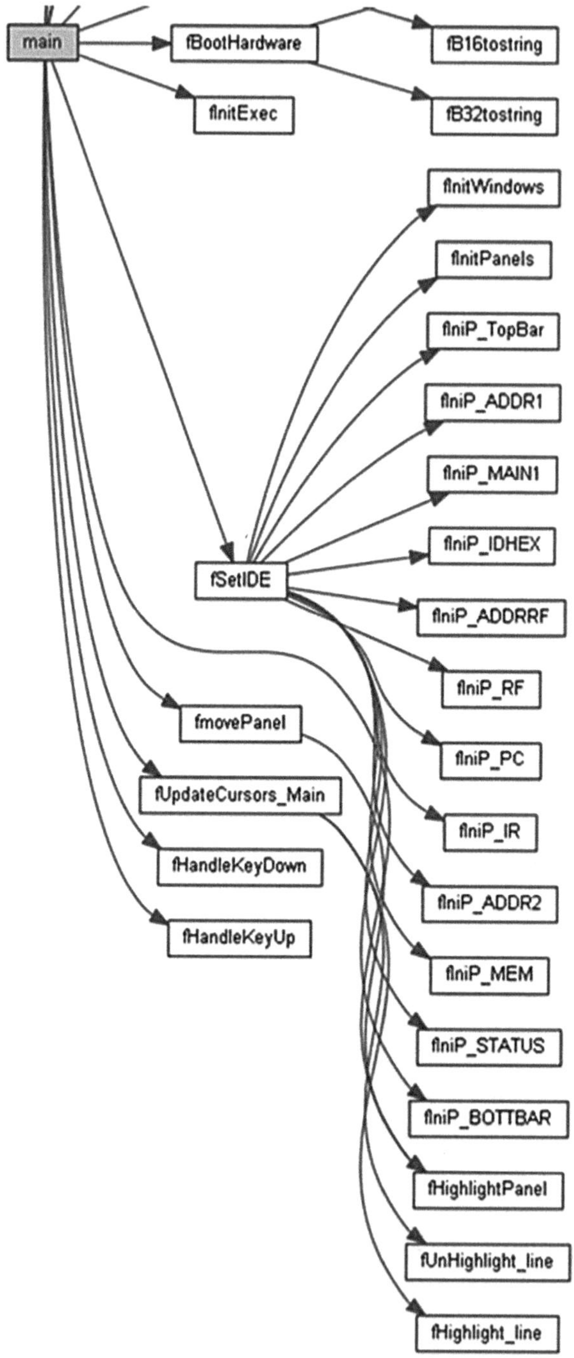

Fig. 8.12 Flow of standard ERA testing (*top*) and testing with Dissimera

8.7.2 *Dissimera Log Sample*

A Dissimera log is divided in four-line units. What follows is an example of a Log File related to the execution of the disassembler. Each unit represents information on 32-bit. The first line of each unit shows the instruction memory address and the hexadecimal value of each 16-bit half. The second line contains the binary values of each half. The third line and fourth lines include information on the assembly instructions of the first and second half respectively, and their pseudo-code meaning.

8.8 Conclusion

- In the development of reliable architectures there is a need for providing accurate testing and debugging of hardware and software. In this chapter we first show the implementation details of hardware architecture that are relevant for simulation.
- We demonstrated ERA instruction execution flow and explained how it can achieve the decoding and execution of two instructions per single fetch and its implications on system compilers.
- We also provide an overview of the hardware prototype and its memory mapping together with an architectural comparison to other relevant RISC and CISC architectures.
- We introduced Minimal Instruction Set Architecture that simplifies the instruction decoder design and the overall system's reliability. ERA's ISA has 16 instructions, does not have a pipeline and provides only absolute memory addressing.
- We argue how far from being a drawback, this simplicity is sufficient to perform safety-critical code improving efficiency reliability.
- We provide the details of a testing and debugging methodology of a hardware prototype.
- Finally we showed the features and implementation of an assembler, and a disassembler/simulator as a proof of concept of the architecture. These custom tools are useful, not only for testing and debugging of the hardware prototype, but for the system and application software.

Chapter 9
Conclusions

9.1 What We Have Done

The main objective of our research was and is to find new ways to introduce new properties into computer architectures and understand drawbacks of existing computer architectures.

At first we review the classic theories of reliability and fault tolerance and find that:

A. the more components a system has, the higher the probability of system failure and
B. the reliability of a system is often dominated by the reliability of its least reliable component.

We conclude that some of the keys to improve reliability would be simplicity of implementation and careful introduction of redundancy.

The terms Reliability, Fault-Tolerance and Dependability do not cover all the attributes and presents properties of safety-critical applications or, after being re-defined over the years, are ambiguous. As a consequence, in Chap. 2 we provide a novel concept of Resilience that encompasses several attributes adapting them to the safety-critical domain.

A resilient system, over a specified time interval, under specified environmental and operating conditions (performability), "must be ready" (in terms of availability) to perform its intended function (reliability), guaranteeing the absence of improper system alterations (integrity). It must have the ability to conduct servicing and inspections (testability) so that in case of failure quick restoration to a specified working condition must be achieved (maintainability) can be provided or can discontinue its operation in a safe way (safety).

Furthermore, a resilient system must have the ability to anticipate changes and evolve (evolvability) while executing (adaptability), successfully accommodating changes by reconfiguring elements of the system if necessary (reconfiguration).

V. Castano, I. Schagaev, *Resilient Computer System Design*,
DOI 10.1007/978-3-319-15069-7_9

Since one of the keys to improve resilience was the careful choice of redundancy and the manner in which this should be applied, we decided to review the different types of redundancy and how such redundancy is translated into functional mechanisms to either avoid or tolerate faults.

We also provide a full classification of fault-tolerant mechanisms based on the type of redundancy employed and study their benefits and drawbacks. Fault avoidance techniques do not guarantee complete removal of faults and present drawbacks such as cost, speed of operation and chip's area. Therefore, fault tolerance mechanisms are needed to further improve the resilience of safety-critical systems.

In order to select a specific set of redundancy types and effective techniques for their applications in the process of implementation of FT we should first define the different requirements of the particular application. Once the domain and requirements are defined we should select the techniques that are more suitable for such requirements and the level at which the redundancy should be applied.

We realised that further improvement of existing mechanisms is based on understanding and analysis how failures originate, what causes them, under what circumstances, in which contexts and how often they happen. In Chap. 2 we introduce the concept of vicious cycle that explains our interpretation of the reasons behind the performance and reliability problems that jeopardise the continuation of Moore's Law. We also review the fault-failure-life cycle and defined the necessary concepts of fault, error, failure and catastrophic failure.

Since the majority of hardware faults in current electronics are induced by ionizing radiation, we studied the damage mechanisms at the physical level, the sources of error and the micro- and macro-effects of such mechanisms. As a result Chap. 4 provides an extensive taxonomy of radiation effects describing their nature, type of degradation, susceptibility, fault rate trends and recoverability.

From the study of this taxonomy we conclude that as we moved to denser technologies at lower voltages, system SER will continue to rise and in particular the contribution of SEU, SET, MBU and SEFI will increase. We also conclude that current mitigation techniques are not efficient when dealing with certain types of SEE and/or with the upcoming rates.

In Chap. 5 we explain how any fault tolerant system involves a Model of the System, a Model of Faults and a Model of Fault Tolerance. Consequently, we add value to such system by developing a comprehensive Fault Model suggesting methods for recognition and reaction against faults.

We discuss fault manifestation, detectability, diagnosability and recoverability and propose adequate solutions for diagnosis and recovery. We have introduced the principle of reconfiguration of the system and how this might be used for various purposes: performance, reliability and energy wise gain, improving the efficiency of resilience. In addition, we introduce GAFT and extend it by providing the different states and actions required to achieve fault tolerance and therefore improve system resilience.

In Chap. 6, using know-how and conclusions acquired in the previous chapters we introduce a hybrid HW-SSW co-design approach of a resilient architecture with the ability to reconfigure, achieving various levels of dependability in different environments. As part of the architecture, we first introduce the syndrome as a tool to handle new property of the system and analysed it as a process and as a tool for reconfiguration that can provide efficiency of reliability, performance and power consumption.

Any theory needs practical confirmation to define applicability and efficiency limits. We, therefore also introduced the embedded recoverable reduced instruction microprocessor and the ERA architecture defining their active, passive and interfacing zones of information processing.

We keep the redundancy level needed to implement fault tolerance, as low as possible.

With regard to the active zone, the instruction set and its implementation are reduced to the minimum; coprocessors, pipelining and floating-point units are removed which simplifies the processor design and reduces the complexity and fault rates.

We explain the checking schemes and re-execution of instruction mechanisms within ERRIC and how they can improve reliability.

With regard to the Interfacing zone, we introduce the T-Logic as basic unit of reconfiguration and discuss its various configurations.

We introduce the syndrome and explained implementation details and how, in combination with a Memory Management Unit and a Reconfigurable Memory Scheme, it can act as a control centre of three functions: fault monitoring, reconfigurability and recovery.

As part of the passive zone, the reconfigurable memory scheme can operate 25 memory configurations and support graceful degradation.

We quantify the probability of state transitions and provide a Markov model of reliability for ERA's configurable memory. Finally in this segment of research we describe the system software support for testing and reconfiguration. We show that by combining this novel hardware architecture with the system software, all key properties of performance, reliability and energy-wise functioning could be improved.

In further chapters we provide the implementation details of ERA's hardware prototype. Having a software simulator of a hardware platform at hand is very useful to speed up software development and debugging of applications.

We developed an accurate hardware simulator with graphical user front end called Dissimera. Dissimera's main goal was not speed but to simulate reliably and accurately the basic core features of ERA with fully reproducible results.

The simulator is built extendable; once core simulation is achieved, we will escalate from there adding new features with an agile methodology. The development of such disassembler/simulator gives us the possibility of:

1. Testing and locating errors of design of the soft core processor;
2. Understand the smallest details of the ERRIC functionality;
3. Simulation of the current version of the processor and the FT version of the processor;
4. Testing and debugging of errors in application and system software.

Finally, we introduce a testing framework that in combination with Dissimera's, with ERA's assembler and with commercial hardware simulators can properly test and debug not only the ERA's hardware prototype but ERRIC's application and system software.

9.2 Next Steps

Arithmetic and logic units are both implemented through the use of logic components. It is known that an arithmetic instruction can be translated into several logic operations. Applying this principle, if an arithmetic unit is suspected of not being able to provide correct service, arithmetic instructions can be translated into logic ones that can be executed by the logic unit of the ALU.

Further research could be done on determining if logic operations can be translated into a set of arithmetic ones and how can this be implemented. What would it be the complexity of such translation? Performance would be affected (graceful degradation), but this technique would allow a running program to finish before recovery or fail-safe restart takes place.

The impact of the size of the register file on overall performance of processor is also a question of further research as in Sect. 6.2.1.

Regarding Dissimera, although basic functionality has been achieved, the implementation of Dissimera is still a working progress:

- The design is completed and the user interface is fully defined.
- Assembling of pseudo code using ERA assembler-preparator (100 % completed).
- Disassembling of binary into human readable code (assembly code) (100 % completed).
- The simulator is capable of parsing the binary file resulting from the previous step and is then capable of classifying data and instructions (100 % completed).
- Simulation of main memory, register file, program counter and instruction registry is almost completed (90 %).

In future revisions we have the intention to include support for: syndrome, extra memory configurations including 16-bit memory configurations and fault-injection. Also:

- We are very interested in finding ways to exploit the functionality that the syndrome can provide.
- We believe that for safety-critical missions such as embedded systems in satellites or space further research is needed.
- We would like to pursue more research in dependency matrix mapping of symptoms and failure modes.
- We would like to apply the context sensing (e.g. altitude, latitude, temperature, dynamic events such as solar flares and weather forecasting) and experience to system software in combination with the syndrome to be able to see and prove the level of resilience of proposed approaches and our architecture.

Chapter 10
Vision on Evolving System Future

10.1 Fundamental Problem

A sequential way of thinking developed by the human race through evolution was constructed on a framework of languages and accompanied grammars (Williams 2003), where differences in dictionaries, alphabets and grammatical rules were visible, but not critical (Schagaev et al. 2013). Indeed, we speak, write and listen to each other sequentially—word after word, phrase after phrase—thereby creating a *one-dimensional* sequence of information.

With the appearance of technological support for logical and arithmetic calculations, computers implemented a system of processing which mimicked our habit of "sequentialism". Calculations were implemented on developed hardware.

Modern hardware technology squeezes logic elements down to the nano-sized with changed hardware structure, mapping electronics into a matrix of interconnected blocks—in other words, hardware topology became *two-dimensional*.

There have been recent attempt to make 3D chips of memory and processors—a simple Internet search indicates (40 million hits on the subject) that since 2007, IBM, Intel, Samsung, Toshiba and others are progressing towards the production of three-dimensional chips.

There are declarations that Moore's law will finally be overcome and with the corresponding boost in performance for future computers being substantial.

It is worth mentioning here that "Moore's law" actually is not a law of nature, or physics, thus a discussion of any limitations to "Moore's law" is in some sense irrelevant in the first place.

This is a subject for research regarding the impact of mass media on technological developments; social science researchers might find it interesting.

There is an unsettling feeling however that the professed goals will not be achieved. This feeling is based on a simple model: imagine a rope, of substantial length, which we need to put into a square box of much smaller size.

V. Castano, I. Schagaev, *Resilient Computer System Design*,
DOI 10.1007/978-3-319-15069-7_10

There is no doubt that this process will take time and the boxed rope will not be as convenient to use as the straight one.

Now imagine that the box become three dimensional; it is debatable whether our ability to use rope efficiently will grow at all . . ..

Adding the need to accept dynamic changes to some segments of rope—updating, or deleting them and then reassembling the rope as a whole, typical functions of run-time system—can reduce any optimism about the impact of 3D to next-to-none. "Rope" here means, of course, a generalization of a program with code and required data.

Figure 10.1 illustrates some of the problems we are facing in computer design. It is clear that the marriage of one-dimensional programs with two- or three-dimensional hardware may not be a successful one. It is also clear that the inclusion of dynamic support for this collaboration will cause even more complications.

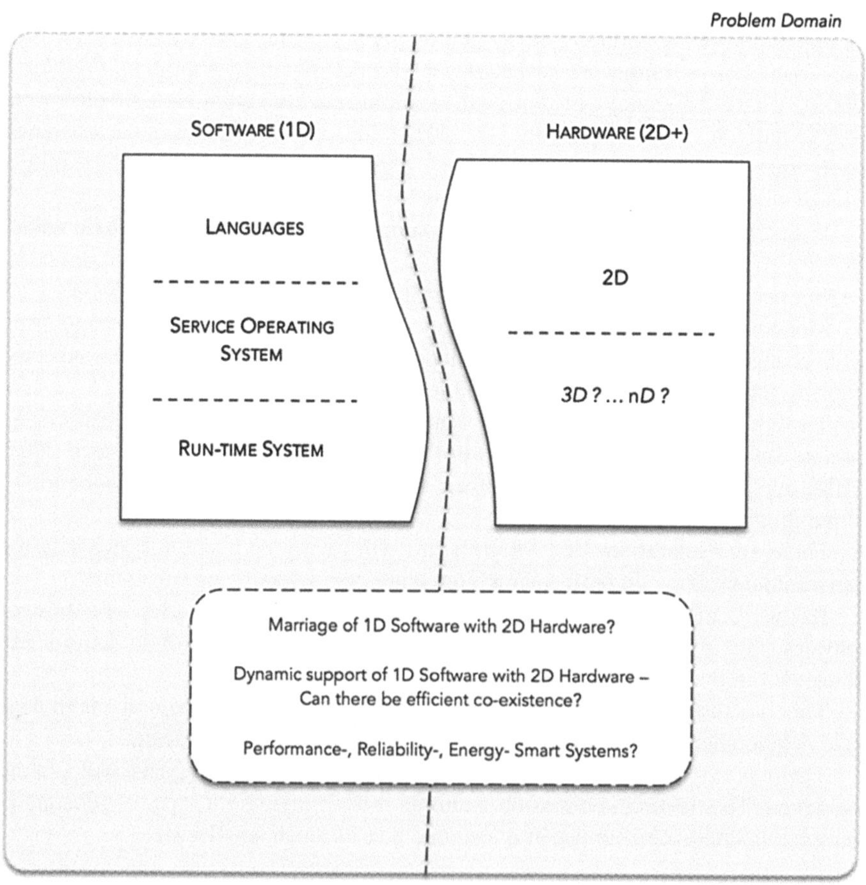

Fig. 10.1 Computer design problems

10.2 Known Solutions (What We Have . . .)

Computer science discovered parallel computing in the late 1940s to early 1950s (Goldstine and Goldstine 1946), (Everett and Swain 1947), (Whirlwind 1952), (Smithsonian 1990), (Hofstra 1999), (Holland 1960), (Newell 1960), (Schwartz 1961), (Slotnick et al. 1962), (Squire and Palais 1963), (Gosden 1966).

The real boom started just a few years ago. An absolutely stunning level of technology and hardware density has been recently achieved (Nair 2002), (LaPedus and EETimes 2003) with processor frequencies up to 4.7 GHz and beyond (IBM 2007).

Surprisingly, the main questions in computer design and technology are still the same as half a century ago. Specifically, the most challenging goal is the design and development of a system that properly uses the relative dependencies of performance, power consumption, cost, reliability, adaptability and flexibility.

None of these dependencies are studied *together* except in Monkman and Schagaev (2013). Nobody even set a goal to make a system that will be efficient and flexible along the lines previously mentioned.

It is clear that we need to describe the relationship between performance, reliability and energy-efficiency of architecture, ability to "trade-off" these properties through design, and observing how these properties change over the life cycle. It was only initially described by us in Monkman and Schagaev (2013).

The theory and development of a system with the required properties that consider those properties as dynamic processes, as well as their interdependencies and modifications (evolution), is in its infancy.

What we have to do now in ICT does not look encouraging: in system design, we have failed to achieve even doubled performance by applying a doubled frequency and quadruple energy, we have failed to achieve any visible system performance growth using the same programs.

A part of the problem is of course that computer technological progress was not accompanied, or supported sufficiently, by theoretical development, i.e. something serious was missing in the first place.

Applying our mantra: "*Any theory values by power of prediction and applicability*" we accept that computer science has failed to lead the ICT world and has closed down discussion of technological developments.

In system software, the situation is confusing and again, not overly optimistic: an overcrowded family of languages (2,500 members) (Kinnersley 2009), with slightly fewer run-time systems (500+) (Berka 2009) shows a confusion and absence of breakthrough developments in terms of efficient system software and hardware designs, and coexistence.

The market domination of one or two operating systems, or languages has nothing to do with a "best player wins" rule.

For example, state-of-the-art hardware illustrates a complete loss of direction; Intel's attempt for an 80-core processor system has ended up with statement Ames (Ames 2007):

> ... Despite using such an efficient grid, the researchers found they could actually hurt performance by adding too many cores. Performance scaled up directly from two cores to four, eight and 16, performance began to drop with 32 and 64 cores ...

Examples prove nothing, but it is clear: a new design strategy is required for the next generation of computer architectures. It should start right from the design phase and be pursued through the whole life cycle of the system and applications.

Any new design strategy should have the goal that the system evolve with support from both hardware and system software.

This feature of "evolution" should be available on demand by user applications and operational requirements. It is also clear that the next generation of computer systems have to address all of the three prerequisites for evolution: Performance-, Reliability- and Energy-smart functioning.

They should be pursued from the first phases of system design, enabling an efficient trade between P, R and E, if user or environment require.

10.3 Attempts to Evolve

Our previous attempts to revise existing designs and find balanced developments of ICT hardware and software were presented in Schagaev et al. (2013), Kaegi and Schagaev (2013), Plyaskota and Schagaev (1995), and Monkman and Schagaev (2013). This work summarizes our earlier concepts and developments.

At first, we propose a revision of the area of computer designs and development by introducing a new system paradigm. Any paradigm has to be useful—see our mantra above.

Thus, we have to develop supporting theoretical models and prototype system in both software and hardware. Several holistic principles we have been using, are as follows:

- Simplicity and redundancy.
- Reconfiguration and scalability.
- Reliability and fault tolerance.
- Energy-wise design.

These principles are explained to some extent in Table 10.1.

As described in Kaegi and Schagaev (2013), the declared principles have to be pursued by applying redundancy in the form of *information*, *structure* and *time*.

The whole process has to be revised from concept through development and the construction of algorithms, the further joint design of hardware and system software and implementation and maintenance, each time pursuing a goal of simplicity and introducing essential redundancy.

This would define the ability of future systems to reconfigure themselves for the purposes of Performance-, Reliability- and Energy-smart operation.

Table 10.1 Holistic principles of computer system design

Simplicity	Complexity is difficult to implement and handle efficiently. To reduce complexity, "bells and whistles" in the architecture are excluded, together with compatibility with main market players, or conventions (pipelines, caches, TLB, etc.).
Redundancy	The deliberate introduction of hardware and system software redundancy and monitoring schemes provides the required level of reconfigurability to reach performance and reliability goals. Redundancy has to implement three processes: checking, recovery preparation and recovery.
Reconfigurability	Reconfigurability serves three main purposes: performance, reliability and power awareness. Handling reconfigurability should be done using Graph Logic Model and run-time support. Active support for reconfiguration during design and at run-time by hardware and system software resources is required for scaling the system up to 100+ processors.
Reliability	Minimum hardware redundancy for design of highly reliable main components. Redundancy of system software and hardware maximizes both tolerance to malfunctions and permanent faults. System software, therefore, supports hardware fault tolerance.
Resource-awareness	Mission critical systems have significant limitations in terms of hardware resources and power consumption constraints, (e.g. battery life). Thus, for wise resource use, reconfigurability must be introduced.

It means also that new systems have to provide an efficient process for "trading off" reliability, performance and power consumption.

Reconfigurability might be implemented efficiently only when hardware designs are supported by system software solutions.

Semantics and the structures of computer hardware vary. The area where information is transformed (we call it the Active zone) is complex and largely irregular. The Passive zone of hardware—where information is stored—has a regular structure and the highest density of transistors.

An Interfacing zone acts as a bridge between the two and serves to maximize the speed of data exchange.

Table 10.2 describes how these principles can be supported and followed during the development of new hardware architectures (HW) and system software (SSW).

When used in combination, it might allow the next generation of computers to be free from existing limitations.

In terms of information processing, computer architectures can be divided into three areas: (1) the processing area—further called the Active zone, (2) the storage area—called the Passive zone and (3) the Interfacing zone. All the three zones have to be reconfigurable for their own purposes and the needs of other zones.

Dependencies between zones define the level of reconfigurability and flexibility of the hardware and system software parts of the system architecture. Note here also, that P, R and E are no longer system requirements, as they become *modifiable* during life cycle: from the first phases of system design (Plyaskota and Schagaev 1995) down to maintenance.

Table 10.2 Properties vs. components of the system

		Simplicity	Redundancy	Reconfigurability
Hardware zones	Active (processors)	Only essential for instruction execution hardware should be developed, duplicated instruction "for convenience" or compatibility should be excluded; no not necessary back-sire actions of instruction execution, such as signs and flags.	Extra elements and schemes should be added either for improvement of reliability or energy-wise functioning.	Designed processing element should be internally reconfigurable (not used hardware is disabled) and support of reconfigurability of the system.
	Interactive (internal and external interface)	Minimum hardware with when possible state-less designs should be used, using asynchronous logic as much as possible.	Designed interconnections must be identical in terms of functioning, making maximum reconfigurability possible, able to be used as direct access inward and onward.	Control of physical and virtual addresses.
	Passive (memoires)	Reduced complexity of address schemes (internal and external).	Virtual memory (of the same type).	Ability to switch data and program areas, hierarchy of recovery points.
System software	Language	At language level—cancelling of complex data structures; limitation and simplification of program descriptions and code structure.	Generation of more than one code to run a program for various requirements: P.R.E Automatic hierarchy formation of recovery point (also for reliability of operation).	Semantic modification of program (graph-logic and control–data–predicate models, see further) to reduce dependencies between code segments and data; support at the language and operating systems-service operating system and run-time system), processes.
	Monitor of processes	Simplified support of process synchronization.	Reducing the number of states in synchronization systems and hierarchy of processes; Implementation of universal loop with ability to control state of hardware and system software at the same time, reducing resource wasting.	Implementation and control of software during run-time and hardware—making "integrity map" possible.
	Reliability support	Recovery points of a program, modules and tasks—to implement hot swap (reconfigurability) when needed. Saving only essential data for concrete execution of a program.	Reducing amount of supportive data in run-time system for program monitoring Introducing of reserved data blocks to run hardware with deficiency (for reliability purpose of view).	

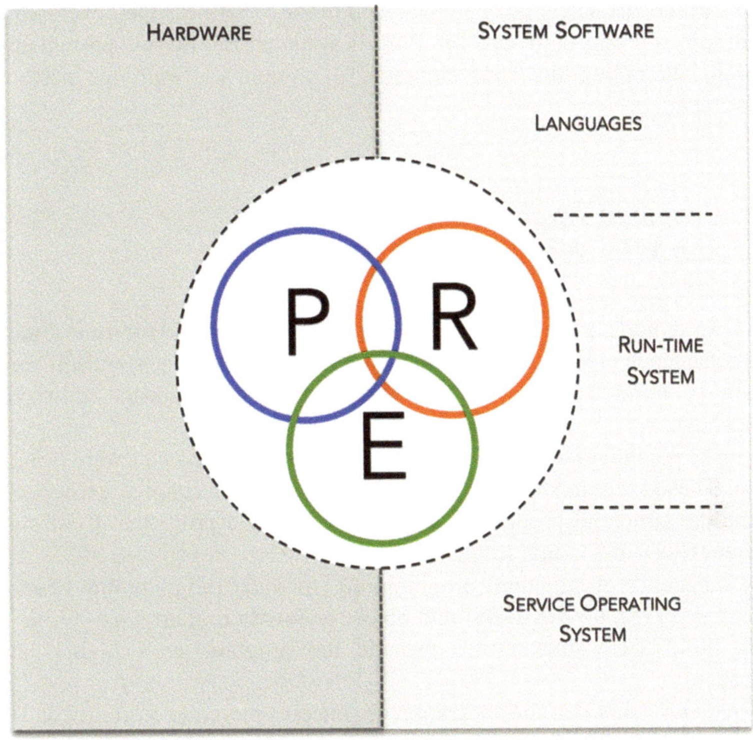

Fig. 10.2 System properties reflection at system software

Therefore, the requirements of evolving systems need themselves to be considered as processes. P–R–E dictates that system software has to be modified, including language, run-time and service operating systems.

Figure 10.2 below qualitatively shows that system Performance (P)-, Reliability (R)- and Energy (E)-smart design might be implemented by means of various methods (circles) and to achieve any development we should choose only not-mutually exclusive solutions.

It is worth mentioning that the properties of system software that are required and useful at program preparation time, are not required at all during program run. Obviously, they should be "stripped off" program code and instead, a "fine tuning" of the code applied to fit existing hardware.

The communication, or interaction between agents, processes, communications, monitors (use any other name from modern avalanche of terms), should also be implemented by taking into account P, R and E requirements, as well.

One of the immediate observations that can be made from Fig. 10.2, regarding performance, is that: the parallelization of a program should be prepared and supported as much as the program structure enables, or even more, while concurrency within the program and during execution should reduced to a minimum, or excluded.

It is beyond the scope of one chapter to describe all possible developments of system software and hardware for P–R–E systems. Below, we present just one approach, attempting the redevelopment of system software for performance improvement.

10.4 Proposed Approach (What We Need and Why We Need This)

At first, we should choose the nearest, "rigorous" language and run-time system and define a sequence of steps that lead to the modification of the system software with the aim to improve reconfigurability and other requirements shown in the Tables above.

To make reconfigurability work, we describe program and hardware using meta-models. These meta-models are independent from the technological aspects, but describe the structural properties of the algorithm and programs: Control–Data–Predicate (CDP) and Graph Logic Model (GLM).

The languages of structural programming, modular programming (Turski and Wasserman 1978), (Wirth 1988) and object oriented programming (Wirth 1989), (Wirth 1992) were invented to provide the programmer a higher level of abstraction.

There is no doubt that the separation of concerns increases visibility and clarity and indeed, eases the process of programming.

These properties, unfortunately, are not necessary and even reduce performance during execution, when hardware load and power consumption are crucial.

In terms of P–R–E, the properties of the system all attempt to deal with parallelism and concurrency of programs and their execution at the same time. They are still far from providing substantial gain, in spite of extreme efforts (Flynn 1972), (Fisher 1983).

Dijkstra (1965) has published an extension of the Dekker algorithm for a solution of the concurrent execution of multiple processes and it is this solution, in one form or another, that is used within languages and hardware (instructions like "test and set", IBM360).

The problem is the automatic extraction of the program parts that are independently executable; parallelism is still a "work in progress" and attracts the attention of researchers on a regular basis, all around the implementations of mutual exclusion and synchronization (Lamport 1983; Lamport and Melliar-Smith 1985), (Gutknecht 2006).

In discussions of performance gain system from software parallelism vast majority of publications are oriented on Amdahl's law.

However, Amdahl's simplification of the efficiency of parallelism by cutting—see our aforementioned rope in n chunks—does not reflect the fact that chunked segments *should be delivered* into the processing areas, and the results *subsequent to* processing *should be collected back*, usually into one block of data.

Read—in Amdahl ratio denominator grows, *reducing* gain from parallelism even more.

Thus, the roles and impacts of language support of parallelization, service operating system and run-time system were largely ignored.

Therefore, predictions of Amdahl about the gain achievable by parallelization were extremely optimistic.

The following works and publications: Nair (2002), LaPedus and EETimes 2003), Hennesy (2008), Bryant and O'Hallaron (2002) highlighted problems of massive parallelism in future hardware designs, showing that the way we approach them either limits seriously or makes impossible to cope with the requirements of configurability and scalability:

> What is good for one purpose is unacceptable for another …

Due to architectural and technological limitations, known hardware designs are not flexible enough in the mapping of algorithms to the hardware, and above all, do not support changing of software structures or hardware configurations "on the fly".

Commercial systems (Windows, Linux, etc.) have shown that attempts in the effective use of hardware features have largely been a failure: the chip frequency now goes beyond 5 GHz, but hardware design and existing software reduces useful system performance down to the 100 s of MHz range.

To cope with this, future architectures must be redesigned with a justification of concepts and arguments from the user level of abstractions, down to the hardware level of representation, aiming at minimizing the degradation of any required characteristics.

Technologically, new hardware chips have seemingly astronomical numbers of logic gates and high physical density. This gives substantial advantage for programming multiple-data problems, where it is possible to use matrix algebra, or multiple data tasks with simple instructions. In turn, the density of hardware elements has reached a technological limit of thermal density.

As already mentioned, accepting the principle of separation of concerns, the programming language has to be hardware independent, as in this way, the logic and complexity of an algorithm will be visible and separated from hardware implementation. From another point of view, we should improve efficiency of hardware and software work.

So the question remains:

> How do we achieve the PRE-smartness of a system by modifying algorithms and implementation?

Certainly, the compilation process of a program should be "fine-tuned" for execution on the specific hardware.

Additionally, however, reconfiguring hardware before program execution also reduces hardware overheads.

Examples prove nothing, unfortunately. In much more general terms, program tuning for hardware Performance- or Reliability- or Energy-smart operation is

Table 10.3 Phases of software development

User oriented phases (UW)			
Concept	Design	Compilation	Implementation
System software and hardware oriented phases			
Analysis of the program through Graph Logic Model to find intrinsic parallelism in Control, Data and Predicate dependencies.	Recompilation of the program into parallel form with introduction of hardware configuration and reconfiguration features into the program, as well as concurrency management - formation of an execution string.	Dual String Execution (New RT data structure that unites arrays and records).	

becoming some kind of reversed programming, where hardware, not the user (programmer), takes the dominant role.

Thus, we might explicitly seek to separate "what is good for the user from what is good for the system and hardware" and, where possible, spread these requirements through the life cycle of software and system development, as Table 10.3 illustrates.

The separation of concerns and goals of PRE-smartness requires supportive models of algorithmic applications from the system software point of view (language and run-time systems) and at the same time, improve hardware efficiency.

Two models we propose for this are called Control–Data–Predicate (CDP) and Graph Logic Model (GLM).

10.5 Supportive Models

10.5.1 Control–Data–Predicate (CDP) Model

Within CDP model each operator in a program can be defined in terms of modifications of three connected graphs: Control (C), Data (D) and Predicates (P), illustrated in Fig. 10.3.

The Control graph presents a sequence of instructions for assembler or processor, or operators for language. Data dependencies related to each operator and data that might be modified along program execution are illustrated by the Data graph.

Finally, a graph of Predicates defines the change of conditions inside hardware and program to enable branching. Thus, every step of program execution is described and defined by a change in three graphs.

While the program operators define a number of elements (nodes) in each graph, the volume of the resulting predicates, state registers, processor and Program Status Words (PSW), physically represented in the hardware by so-called processor status registers, contributes to the "width" of each layer of these graphs.

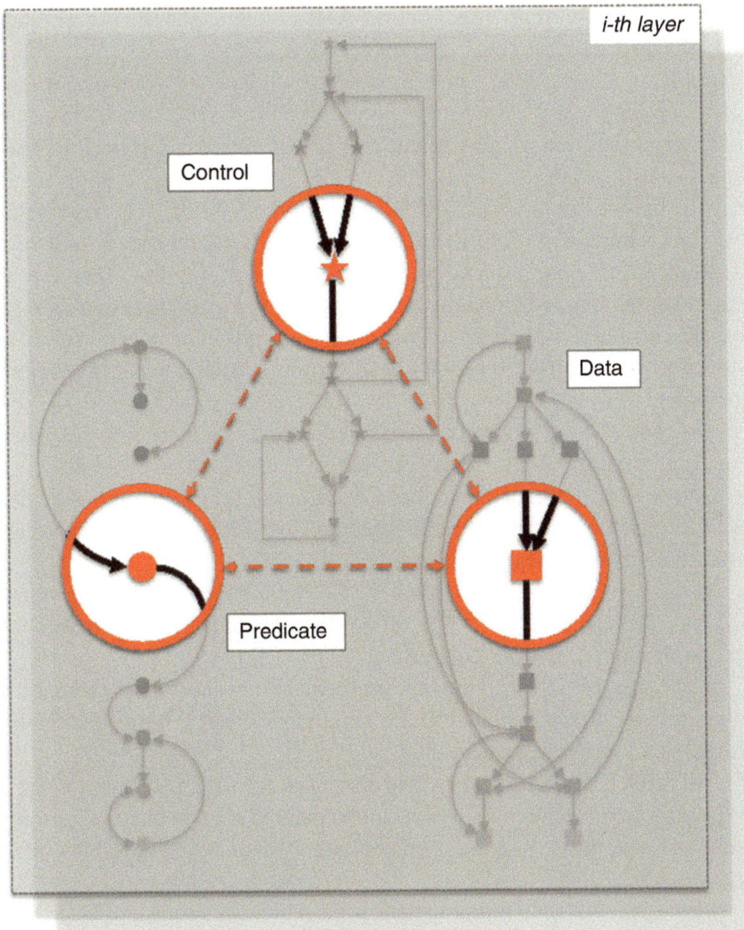

Fig. 10.3 Control–Data–Predicate (CPD) model

To support and exploit parallelism, we should simplify all three graphs by separation of Data and Control, making them as independent as possible.

To support simplicity, a graph of Predicates should be simplified and fragmented—in other words we propose that Predicates should be processed when they are needed for decision-making and not stored in the hardware.

To support reliability and fault tolerance, the detection of possible changes to program state caused by hardware deviation is required, together with the development of recovery schemes of hardware and software (Kaegi and Schagaev 2013).

Therefore, any complex operator and hardware instruction respectively jeopardize the possibility of generating "snapshots" of the previous hardware states, as well as the flexibility of program segmentation, allocation and reconfiguration.

In the vast majority of architectures, as well as languages (their compilers) and run-time systems, the functions of data access and data processing are mixed—tightly coupled—and hardware state modification due to program execution is not controlled explicitly.

This makes condition change latent and execution of a program unjustifiably complex.

In a typical processor, such as ARM, Intel and SPARC, the Arithmetic and Logic Unit (ALU), or even several of them, as well as shifters, registers, internal cache, special registers, pipeline sequencers, etc., are active during the execution of each instruction, sometimes with several data passes within a single instruction.

Thus, the complexity of hardware handling becomes enormous: 75 % of the die size is occupied by translation look-ahead buffers, caches, synchronization logic and pipelining.

However, none of these overheads are required from the point of view of the programming language and program operators.

Any condition of the hardware related to an operator, or instruction representation (the three nodes within the same layer in Fig. 10.3) requires checking. This makes parallelism or reliability much harder to achieve.

The reasoning is different, however: for parallelism, hardware should be designed as "flat as possible", but reliability demands the limitation of fault propagation through hardware schemes.

The complexity of a system and the implementation cost of parallelization, or fault tolerance are directly related to the amount of the resulting modifications of the hardware and program states.

Our CDP model shows that when P is only used for the selection of the program flow, a special operator and instruction can be defined to generate the current value of P and store the result in a register, making system reconfigurability easier to achieve.

To summarize, for the implementation of parallelization at the level of the instruction set, the design objectives will be:

- Implementation of "as simple as possible" logic to form predicates;
- Mapping the language operators as close as possible to the processor's modified instruction set;
- Reducing the size of the state space that needs to be saved before the execution of each instruction.

This brief analysis of CDP shows that clarifying the interaction of language and hardware can save us from "improvements" that drive mutual loss. Studying the CDP scheme of a program also allows potential program parallelism to be checked in all three, graph dependencies. However, this is not the whole picture.

10.5.2 Graph Logic Model (GLM)

As a further improvement in parallelization and at the same time concurrency reduction, we need to handle the actions that were initiated in parallel, but eventually end up in conflict when one or more resources are accessed.

To describe this, we introduce the Graph Logic Model (GLM), Figure 10.4. Every meta-program structure might be described using GLM.

GLM provides a scheme to redevelop existing programs into their maximum parallel and minimum concurrent forms, limited only by available hardware resources.

An indicative example how the GLM works, using a program control graph, is presented in Fig. 10.4.

Note that a GLM might be applied for any of graph of the CDP model. GLM uses logical operators from the set {AND, OR, XOR} for every program, or hardware scheme that it describes. These operators are allocated for the input and output of each vertex.

A vertex might be an operator, instruction, or state. Vertex **a** of Fig. 10.4 may be described thus:

$$\mathbf{a} : \mathrm{OR} - (\alpha\,\mathbf{b}, \gamma\,\mathbf{d}), \quad \mathrm{AND} + (\beta\,\mathbf{b}, \delta\,\mathbf{c}) \tag{10.1}$$

where "−" stands for every logical operator of an output link and "+" for every input link, while α, γ, β and δ are weights, or priorities assigned for the link.

Until now research in parallelism was mostly targeted at finding parallel branches of programs and independent data elements.

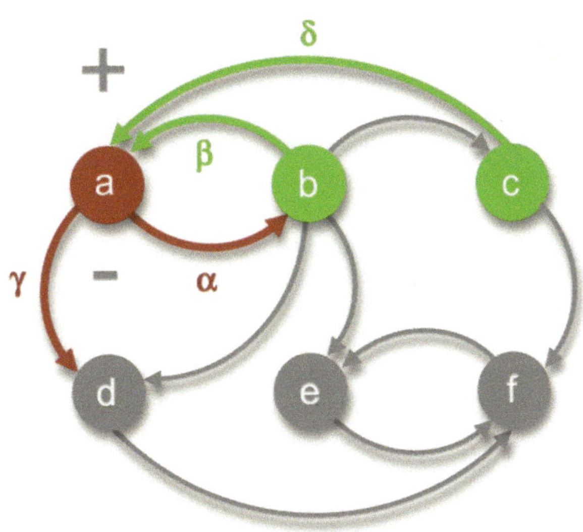

Fig. 10.4 Graph Logic Model

However, expecting pure parallelism is hardly feasible; what is initiated as parallel segments ends up ultimately in concurrent mode, competing for a resource such as a socket, printer or data concentrator.

The rare exception, such as graphic processors with high numbers of SIMD-like processors, just proves the rule.

A simple notation, similar to Eq. (10.1), can describe program structures and hardware structures consistently in terms of coexisting concurrency and parallelism.

GLM *explicitly* separates parallel and concurrent elements in the system description by introducing logic operators into the program graph for incoming and outgoing ends of edges.

The application of the logic operator XOR (exclusive OR) on an input, or output of an edge, defines ALL possible concurrencies in the program graphs.

In turn, all possible parallelism in the Control graph are defined by finding all outgoing, or incoming edges explicitly described by the AND operator. The same approach might be applied for the Data and Predicate dependency graphs of the CDP model.

Moving forward, obvious questions of PRE-smartness arise:

> Can we prove that the hardware representation for the chosen algorithm is parallelized in its maximum possible way?
>
> How does the current hardware–software solution correspond with a limitation of available resources?

The answer to these questions is a part of the design challenge for any evolving system. Having a correct program, CDP and GLM can then be applied to extract the parallel segments and data paths and help in reducing concurrency.

This reversed programming gives us a chance to play with the software and monitor system software redundancy (deliberately introduced at the recompilation phase) and hardware redundancy (introduced at the design phase) on the fly.

It also enables to use reconfigurability of the system for Performance-, Reliability- or Energy-smart functioning when it is necessary.

Figure 10.5 introduces the use of CDP and GLM for program parallelization. There is no doubt that we have to start from correct sequential program.

The proposed sequence clearly separates properties required at the *program-writing phase* from the properties required *during execution*, as the latter are dependent on the hardware resources that are available and may change during execution.

In this way, hardware features (that might be dynamically adapted during run-time) become represented in the program logic.

A new language is needed to naturally express algorithms in a form that supports program recompilation into a redundant form fitting with the "hardware view".

This does not mean that other programming paradigms such as structured, or OO are no longer useful. It means, however, that they have their limits and mostly serve user convenience and interfacing.

The newly proposed paradigm serves to improve performance of the end product. This is achieved with the design of a new reconfigurable architecture, a system software and the dynamic transformation of existing software into a form that is

1. Debug sequential algorithm (SA)
2. Create a model of the SA as triple <C, D, P> where C, D, P are graphs of control, data and predicate dependencies of SA
3. Apply the graph logic model to modify C, D
4. Where possible substitute "strong" logic operators by "weak" ones. An "IF" might be substituted by XOR (to start) and OR, or AND or NOP, leaving an option to parallelize segments of the algorithm
5. Check all introduced XOR in sequence for potential concurrency in the algorithm – for the OS

Fig. 10.5 Sequence of steps to make sequential algorithm parallel

efficiently executable by adaptive hardware: Evolving Reconfigurable Architecture (ERA), or more generally, Evolving System (ES).

Interestingly, the algorithm of parallelization might be applied for all three dependency graphs—Control, Data and Predicate—and, therefore, deduce the maximal parallel form. The availability of hardware serves as the termination condition for this algorithm.

10.6 System Software for Evolving Systems

System software for the next generation of evolving systems includes both the design of a new modular programming language called Active Language (AL), and two operating systems: a run-time system, called the Active Reconfigurable Run-Time System (ARRUS) and a service operating system (SOS). AL and ARRUS are tightly coupled, as AL needs run-time system support for some of its features, such as recovery.

10.6.1 Active Language (AL)

The language has to be able to describe active processes and is, in fact, a direct derivative of Oberon (Wirth 1988; 1992) which includes the following new features:

- A new data structure is introduced that eases the mapping of data to memory (dual string model);
- GLM extensions will be exploited in the control and operator model;

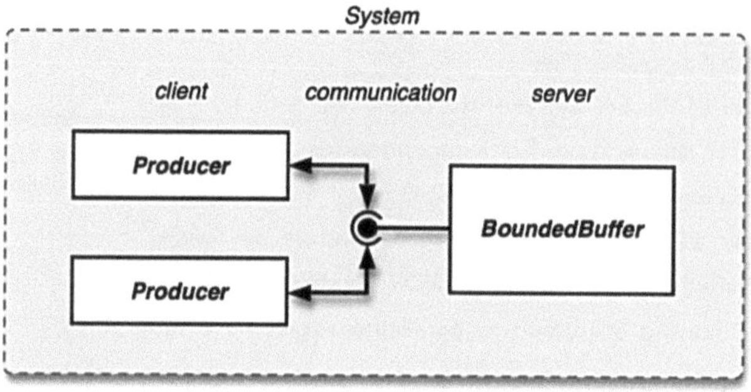


```
INTERFACE DataAcceptor;
[ IN Element (x: INTEGER)]
IN Finish
END DataAcceptor;
```

Fig. 10.6 Describing hierarchical component structures

- Separation of interface and implementation to support dynamic software and hardware reconfiguration;
- Deliberate reduction of the number of supported data structures and other language features;
- Physical separation of constant, global and local variables and introduction of recovery points (Kinnersley 2009);
- Support for recovery and reconfiguration points at the module level using special program structures;
- A special language feature that allows safe hardware access will be introduced which results in a completely type-safe language, without loopholes;
- AL carefully revises dangerous language constructs, such as unbounded loops, and introduces the calculation of upper execution times and stack sizes to ease certification;
- Rigorous memory management strives for an implementation without pointers and references.

An essential set of the above-mentioned concepts has already been implemented in our component-based, parallel programming language COMPOSITA (Bläser 2006), (Bläser 2007).

AL is a further development of this language, synthesizing additional concepts for real-time and fault-tolerance. In COMPOSITA, programs are entirely composed of active components which govern strict encapsulation and dynamic wiring, with a dual concept of offered and required interfaces and communication-based interactions (Fig. 10.6).

As the language is based on hierarchical composition and does not employ any ordinary pointers or references, surrounding components properly control the deletion of components and no garbage collection is needed for safe memory management (Bläser 2006, 2007).

In turn, Fig. 10.7 shows an example of a component structure, where a component can contain an inner network of sub-components. Communications follow a formal protocol written in an EBNF-like notation.

Due to the strict encapsulation, the components and can be easily mapped to various hardware architectures. Inherently, it enables parallelism (N components

```
COMPONENT System;
 VARIABLE
  buffer: BoundedBuffer; producer[i: INTEGER]: Producer;
  consumer[i: INTEGER]: Consumer; i, n, m: INTEGER;
 BEGIN
  (* read n and m from user *)
  NEW(buffer);
  FOR i := 1 TO n DO
   NEW(producer[i]); CONNECT(DataAcceptor(producer[i]), buffer)
  END;
  FOR i := 1 TO m DO
   NEW(consumer[i]); CONNECT(DataSource(consumer[i], buffer)
  END;
END System;
```

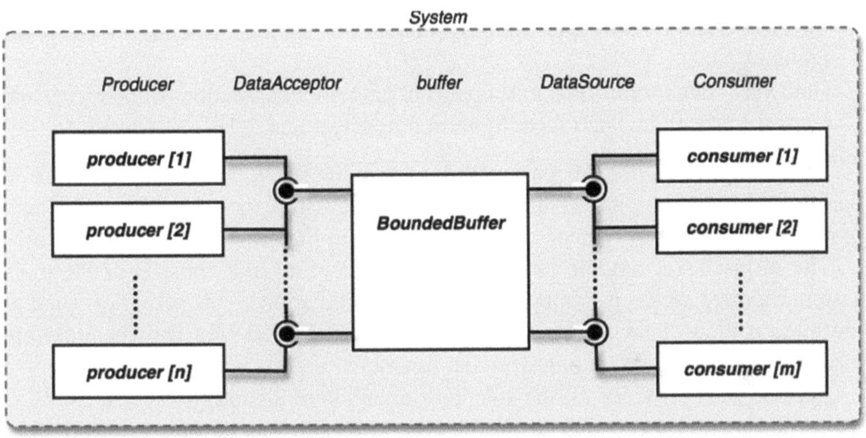

Fig. 10.7 Communication-based component

may be scheduled on up to N processors), as well as redundancy (the same components may be executed as multiple replicated instances).

In subsequent work, a prototype compiler and run-time system for evolving systems has already been developed.

Direct support for reconfigurability and recoverability of program structures at the language level makes reconfiguration of system possible in case of hardware degradation due to faults, or task special requirements of system power-saving operation, or on the opposite, boosting task by using maximum hardware resources for completion of a task with required time limits.

For hardware fault tolerance, where malfunctions outweigh permanent faults, system software at the language level should include recovery points for the program at various levels of program presentation: procedure, module and task.

A detailed description of language support of hardware fault tolerance using recovery points is presented in Kaegi and Schagaev (2013).

Note here that the recovery point scheme will be embedded in the language and oriented on the programs and data structure. This reduces the overhead for recovery after malfunctions and eases the impact of possible permanent faults.

10.6.2 Active Reconfigurable Run-Time System

The run-time system, ARRUS, should also be involved in supporting real-time processing, as well as real-time reconfiguration of the underlying hardware elements and the respective network topology, including:

- Flexible dynamic hardware resource management;
- Software reconfiguration to adapt to changes in hardware states and system conditions;
- Management of hardware/software interactions in the presence of hardware faults;
- Hardware state monitoring and support of graceful degradation using testing and recovery procedures and reconfiguration management.

To match the required reconfigurability, parallelization, real-time, resilience to hardware degradation, and distributed control processing, the ARRUS itself is built in a strict hierarchical manner, as it is illustrated in Fig. 10.8.

The lowest level module has no dependencies at all and consists of the main system monitor which is responsible for the coordination of all activities, such as the initialization of reconfiguration entities, timer services (not shown), interrupt handling and all the remaining depicted functions.

ARRUS also provides all the standard functions of a run-time system, such as memory management, which are well known and explained in literature [34], [35].

These features are omitted in Fig. 10.8 to keep the diagram understandable. In a standard control loop system, it is up to the programmer and the applications to diagnose faults and react appropriately.

In ARRUS, however, this is not the responsibility of the application, but of the run-time system. Thus, ARRUS is responsible for diagnosing faults (software failure, malfunction, permanent fault, etc.) and notifying the appropriate software and hardware monitoring services of any required changes.

In ARRUS, fault diagnosis (software failure, malfunction, permanent fault, etc.) and handling (fault elimination, recovery and reconfiguration) is in the responsibility of the run-time system, not of the application.

User applications are not allowed to communicate directly with the reconfiguration mechanisms. The rationale behind this principle is the idea that the run-time system is the only entity that knows the current hardware state, all ongoing processes and their resources.

In case of a fault, it can thus based on the available resources reconfigure the applications. The following developments will be crucial for the ARRUS fault handling mechanism:

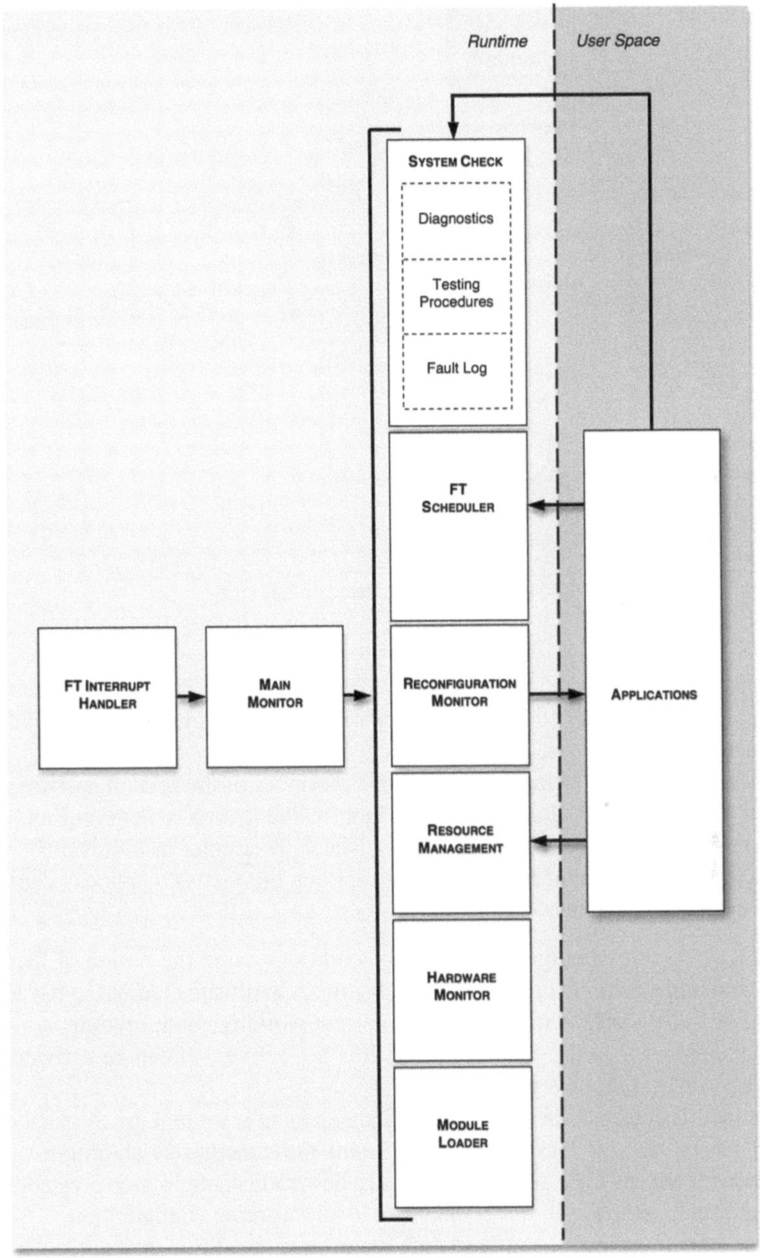

Fig. 10.8 Structure of run-time system for an evolving system

- Monitor hardware state. Possibly, either with interrupts (hardware signals changes), or periodic, software-initiated, hardware checking (further, both approaches are required to implement hierarchical hardware checking) (Kaegi and Schagaev 2013);

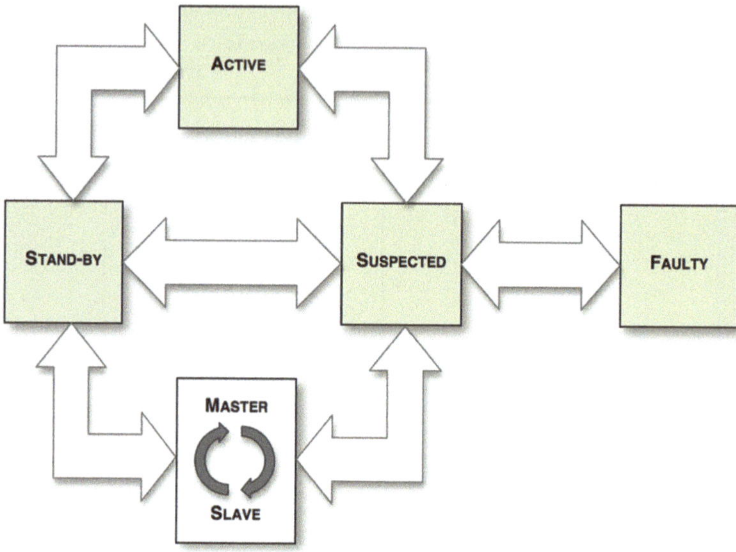

Fig. 10.9 Hardware states from the run-time system point of view

- Reaction on hardware state changes. The run-time system is responsible for the management of the hardware states and reconfigures the applications accordingly;
- Collaboration of checking and recovery processes at the system software level and hardware level. Introduction of fault-resilient, task scheduling and hardware/software fault-handling strategies (schedule simpler task versions);
- Fault tolerant semaphores: A new concept that eliminates deadlocks caused by hardware deficiencies.

ARRUS handles changes to hardware conditions using the notion of hardware states and transitions, as Fig. 10.9 shows below. A hardware element in the Active, Master or Slave states is included in the current working configuration.

If an element is in the Stand-by state, it is not active, but can be activated in a further reconfiguration step.

If a fault is detected, the affected hardware element is set into the Suspected state which means that, at least for a while (until the generalized algorithm of fault tolerance (Schagaev 2008) is completed and a new valid configuration is established), this hardware element will not be included in any working configuration.

This representation of the hardware state for every ERA element defines the current configuration of the ERA hardware at the run-time system level.

A configuration change can be triggered by changed application, power, reliability, or performance requirements, or a detected error.

10.7 Evolving System: Hardware

10.7.1 Basic Schemes

The indicative structure of a hardware element for evolving system is presented in Fig. 10.10. Configurators called T-logic provide for flexible use of processor and memory elements in the system configuration related to performance and health conditions: when one processor is dealing with its own program (self testing, autonomous calculations), it disconnects from the other nodes.

The same technique is used to form reconfigurable hardware that is capable of adjusting to program requirements, or react to other events such as detected permanent faults.

The memory configuration of an ERA element (Fig. 10.10) works for both resilience and performance. A special logic scheme, T-logic, configures the memory structure.

When an application requires maximum reliability, the T-logic scheme might configure the memory as a 4- or 3- (shown) unit with voter.

The configurations: two to compare and one spare, or three independent memory elements are possible. The number of memory elements might vary. For recent implementations, four memory elements were proposed.

However, the principles of configurability by using T-Logic elements remain the same.

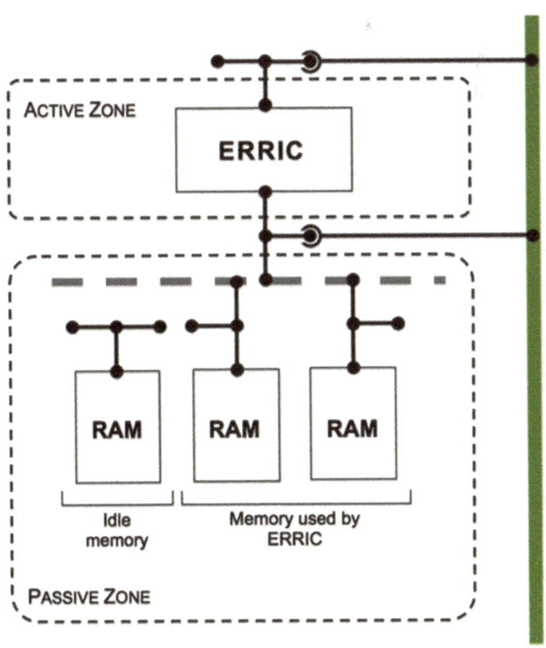

Fig. 10.10 ERA element

The instruction set of an elementary processor is designed to recover from hardware malfunctions by repetition of the instruction, making malfunction tolerance efficient (Schagaev 2008).

In comparison with Motorola, ARM and Intel, the proposed ERA is much simpler, with a higher level of parallelism and frequency able to be achieved. Due to simplicity-by-design, ERA only needs 10 % power, when compared with the competitors, to reach the same clock speed.

The same technique is used to form a hardware configuration that is adjustable to program requirements, or when a hardware element itself (or architecture) detects hardware faults and cannot be involved in further program execution either on a temporary, or permanent basis.

The final decision about permanent isolation of an element is performed during a special mode of self-healing when testing and recovery procedures are executed.

Each element can be turned off individually to decrease power consumption. Note that the structure assumes only one leading element at a time, enforced by a "rotation" of the T-logic element. T-logic makes an ERA possible to operate until "the last man is standing," i.e. until a single processor, called ERRIC, and a single memory element can communicate.

By design, the Evolving, Recoverable, Reduced-Instruction Computer (ERRIC) is able to recover from major malfunctions and repeat the ongoing instruction when an error is detected. The bespoke instruction set and its declared top-down implementation of all required features reduces the impact of malfunctions on the performance and reliability of a system as a whole.

As it was shown in Schagaev (2008), the efficiency of malfunction tolerance depends on the fault coverage and the amount of involved redundancy. The efficiency of malfunction tolerance grows together with the ratio of malfunction to permanent faults.

An indicative performance comparison of Motorola, ARM and Intel to proposed architecture shows that the performance of ERRIC is not clock-to-clock competitive with Intel and ARM, but as the implementation is much (up to six times) simpler, a higher level of parallelism can be anticipated.

Additionally, early power estimation shows that ERRIC needs only 10 % power compared to the competitors to reach the same clock speed. Finally, the absence by design of pipelining support and the ability to access memory at almost half that of, say, ARM, enables system monitoring and processor frequency variance when applications require extreme performance.

Below, Fig. 10.11 shows a first prototype of ERA, designed and developed by the authors in collaboration with ITACS Ltd.

Memory, organized as four, mutually replaceable schemes with virtual addressing configure the ERA for Reliability (two pairs, or triple structure with spare element), Energy-smartness (one element is active, the rest disabled), or full capacity—when all elements form one memory bank for the program.

The ROMs function as one, or both can be active. At the moment, all reconfigurations and ERA processor elements (four of them) are assembled on Altera PLD, shown.

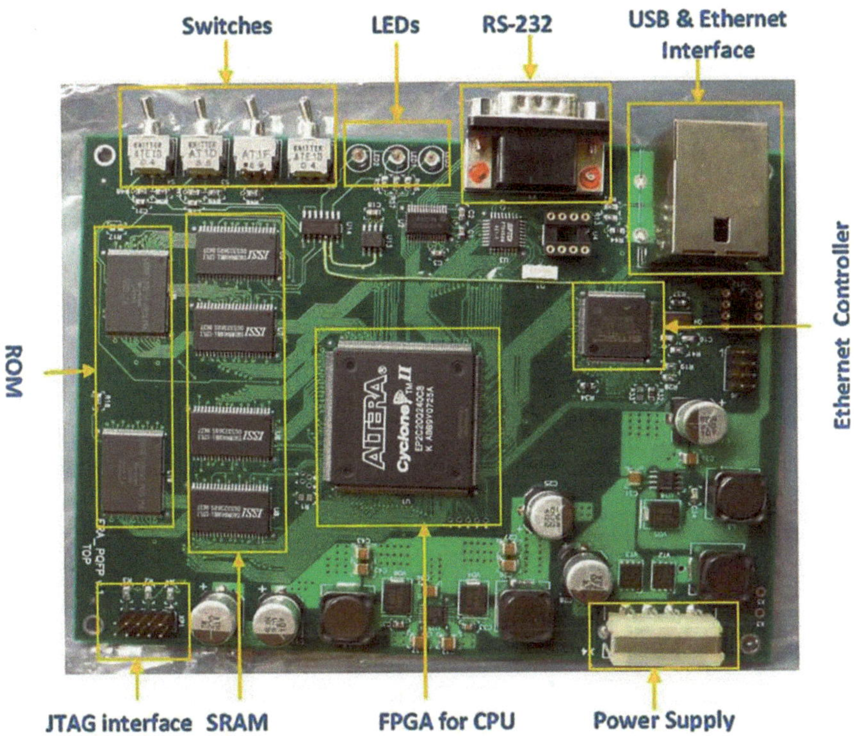

Fig. 10.11 ERA element—first prototype

Further development will obviously lead to special chip for processors, memories with reconfiguration support and specula design of T-logic element to achieve it maximum simplicity, performance and reliability.

The proposed reconfiguration schemes and design approach is the subject of further development, as it provides a unique linear exchange, or reliability/performance and power consumption, dependent on the actual need and request of user task, or run-time system.

10.8 Evolving System: Multi-element Configuration

There is no doubt—see "model of rope and box" above—one has to address how, if possible, we can use the resources of multi-agent systems supported by multi-element architecture.

Conceptually, Fig. 10.12 presents the next step for a prototype of evolving systems.

Each element of the proposed architecture might serve as a node in the system: it can be turned off individually to decrease power consumption, or made inactive yet

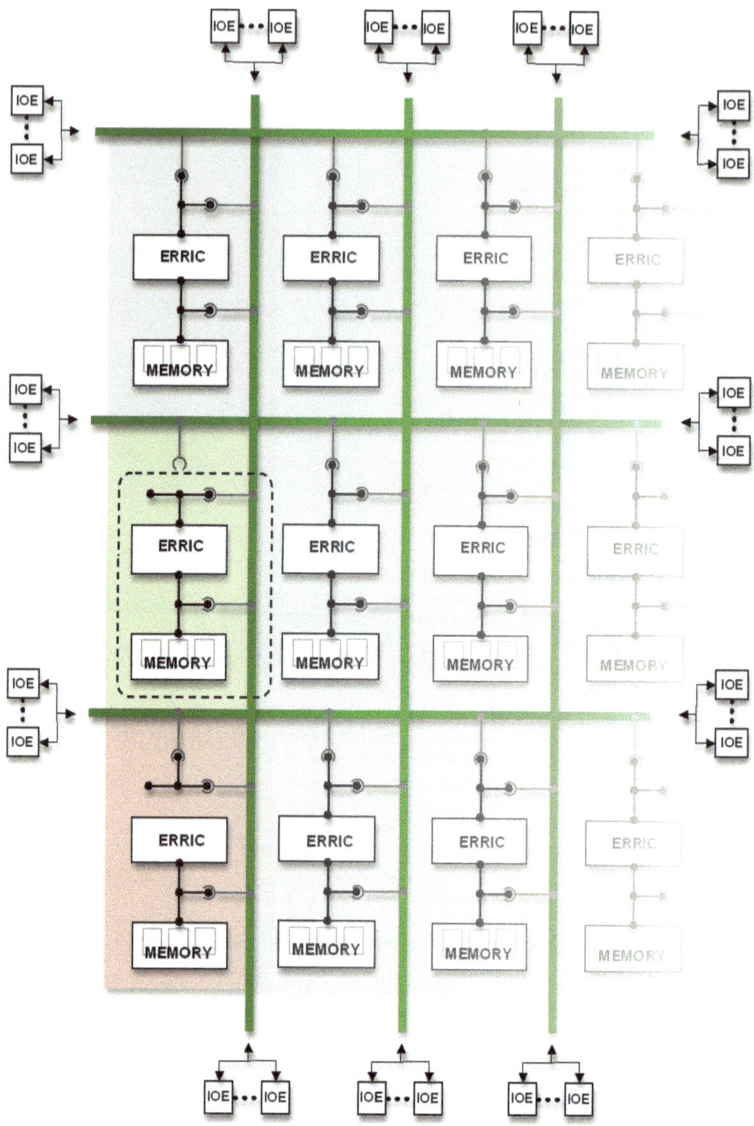

Fig. 10.12 Indicative architecture of evolving system

able to be included in any application configuration, or aggregated with others for maximum capacity or reliability.

To recap, when applied throughout the whole system, T-logic elements enables the whole evolving system to operate "until the last man stands," i.e. until a single processor (on the Fig. 10.10 ERRIC), and a single memory element can communicate through the remaining links.

Note that active and passive elements (processor and memory) can be at the extremities in this scenario.

Various areas of Evolving Architecture can be involved in different types of tasks. Some segments might run safety-critical tasks—extremely demanding in terms of reliability, while others execute heavy calculations of matrix algebra.

The pink-colored part of the scheme illustrates that when, for any reason, the Active zone (ERRIC) is not in action (either because of task configuration, or because a permanent fault has induced its idle status by system software).

Other processors might use local memories of one hardware section within the system.

Finally, the dotted part of scheme illustrates that the leading active element has only one connection, as it is counted as active and initiating an exchange; the others involved are also “listening”.

The proposed approach is promising. Initially it was presented by Prof. Schagaev (London Met) with some assistance from Prof. Gutknecht (ETH Zurich) at EC program FET (Future and Emerging Technologies) Information Day in 2005, January 14th, Brussels.

Since then we have developed: evolving system concept and theory, theoretical analysis of reconfigurability, prototype of system software (language, system run-time system) and hardware, including key elements such as processor, arithmetic unit, logic unit, reconfiguration logic, showing and proving existence of practical alternative to software and hardware brands and market predominated companies.

We do believe that human kind can do much more to propose in this domain of knowledge and technology, leaving Microsoft, Apple, Intel and ARM behind.

10.9 Evolving System Approach vs. Berkley View

The recently published Berkley view on parallel computing (Asanovic et al. 2006) differs from our approach. Table 10.4 is self-explanatory and illustrates the differences, showing both system software and hardware concepts and implementations.

Table 10.4 Berkley view vs. evolving reconfigurability approach

<table>
<tr><th>Berkley view</th><th>Evolving reconfigurability approach</th></tr>
<tr><td>Seven goals (and 11 bullet points)</td><td rowspan="4">New rigorously defined computing paradigm based on GLM exploits automatically maximum possible parallelism of both algorithm and hardware available and minimizes synchronization complexity at the run-time to reduce concurrency. This approach guarantees together with flexibility of hardware configuration for both performance and reliability purposes. Adjust available hardware for maximum possible efficiency across the whole broad class of applications usually covered by semantically different architectures such as VUW, SIMD, and MIMD.</td></tr>
<tr><td>Future computers must be effectively parallel;</td></tr>
<tr><td>It is possible to consider 1,000+ cores;</td></tr>
<tr><td>Performance measurement for parallel computing should be re-evaluated and new metrics introduced.</td></tr>
</table>

(continued)

Table 10.4 (continued)

Berkley view	Evolving reconfigurability approach
Auto-timing of software and hardware;	Software will be written in traditional way, and debugged on a standard systems;
Human centric computing in multi-core;	Parallelization of the algorithm will be done by backward compilation of a sequential program. Using GLM again for representation of task potential parallelism and fine-tuning of hardware resources available on the wafer, before and during application run.
Application of wide range of data typing.	
Three levels of parallelism should be pursued: task level word level and bit level;	Task level parallelism is prerogative of run-time system and heavily dependent on workload and available resources during application run: therefore, dynamic scheduling to optimize task parallelism is exception not the rule in evolving architecture. For evolving reconfigurable architecture dynamic support of parallelism should be an exception not a rule.
Parallel programs must be presented independently to the number of processors available;	Application of GLM for both redesign of algorithms and timing on architecture available resources (supported byT-configurator at the element and architecture levels) maximize parallelism and minimize concurrency.
Limitations on features that reduce parallelism;	
OS functionality should be based on libraries and virtual machines.	
A higher-level rate of fault is expected for multicore systems.	Assumptions about higher rate of permanent hardware faults for the next generation of electronics are not correct. More details see, for example in Feynman lectures…
SEC/DED options proposed.	Number of malfunctions caused externally by alpha particles and internally due to higher density of elements on the wafer does not lead to increased reliability by using SEC/DED; most likely 16+ bit errors will take place in hardware. System software contribution to the malfunction of the system caused by support of dynamic parallelism and complex concurrency monitoring.
Synchronization overheads should be reduced.	Overloaded by task parallelism, the monitoring OS will be deadlocked. We propose a reconfigurability of available hardware and tasks by recompilation of existing algorithms.
Wide range of data types should be implemented:	We propose Dual string data structure where for each data element of the array a special descriptor defines data type (altogether 232 types);

(continued)

Table 10.4 (continued)

Berkley view	Evolving reconfigurability approach
1 bit (Boolean);	This covers all possible data types user can dream or imagine. Efficiency of access to the proposed data structure is equal to the array access.
8 bits (Integer, ASCII);	
16 bits (Integer, DSP fixed point, Unicode):	
32 bits (Integer, Single-precision FP, Unicode);	
64 bits (Integer, Double-precision FP);	
128 bits (Integer. Quad-Precision FP; Large integer (>128 bits))	
New models should support proven styles of parallelism;	Styles of parallelism are application specific and vary due to technology modification. We propose an auto-tuning of existing programs into their maximum parallel form;
FPGA systems are future HW platforms for multi-core computing.	FPGA technology at the element level and specially designed wafer with prefabricated configuration fabric of active processing element and passive elements enables monitoring architecture for performance, power consumption and reliability.

10.10 Evolving System: Conclusion

- This work presents concepts and design ideas on how to cope with some known drawbacks of computer architectures. Evolving systems of future computers will exploit reconfigurability of hardware and software for Performance-, Reliability- or Energy-wise functioning.
- At the system software level, a reconfigurability of hardware and software should be introduced, represented and supported by programming language, service operating system and run-time system. Ultimately, this will result in a simple, yet scalable, reliable system and provide performance and power saving options, with linear trade between.
- To avoid existing drawbacks of parallel and other reconfigurable architectures, several holistic principles were presented and pursued through the whole life cycle of computer systems: from the preparation of algorithms down to the execution of programs and hardware.
- We presented a new approach, aiming to introduce a system with evolving features, based on the mutual design of architecture and system software. System software support for hardware reconfigurability is centered at the compiler level, leaving only essential reconfigurability handling at the run-time level.
- It was shown that consistent support of system reconfigurability eases parallelization, reduces concurrency, assists fault tolerance and implements a power-awareness for applications, when necessary. Two models: Control, Data, Predicate dependencies of the program and the Graph Logic Model represent

concurrency and parallelism of a program and enable their explicit separation. A sequence was proposed to find the minimum form of any program in terms of time overheads, limited only by available hardware resources. The generalized algorithm of parallelization of sequential programs was presented.

- The structure of a reconfigurable, run-time system was discussed, together with a scheme for hardware configurability support.
- Hardware elements and the structure of the whole architecture were presented, with explanations of how maximum reconfigurability ("'til the last man is standing") is achieved.
- A comparison of the evolving reconfigurability approach with known approaches was presented. In contrast with Von Neumann, we are attempting the design of reliable (and reconfigurable) systems with reliability of components.

References

Abramovici M, Breuer MA (1979) On redundancy and fault detection in sequential circuits. IEEE Trans Comput 28:864–865

Abramovici M, Breuer MA, Friedman AD (1994) Digital systems testing & testable design, 1st edn. Wiley-IEEE, Hoboken NJ

Adams JH, Gelman A (1984) The effects of solar flares on single event upset rates. IEEE Trans Nucl Sci 31:1212–1216. doi:10.1109/TNS.1984.4333485

Adams JH, Silberberg R, Tsao CH (1982) Cosmic ray effects on microelectronics. IEEE Trans Nucl Sci 29:169–172. doi:10.1109/TNS.1982.4335821

Agrawal VD, Chakradhar ST (1995) Combinational ATPG theorems for identifying untestable faults in sequential circuits. IEEE Trans Comput Aided Des Integrated Circ Syst 14:1155–1160

Alam M et al (2007) Characterization and estimation of circuit reliability degradation under NBTI using on-line IDDQ measurement. In: Proceedings of design automation conference

Alanen J, Ungar LY (2011) Comparing software design for testability to hardware DFT and BIST. In: 2011 I.E. AUTOTESTCON, pp 272–278

Allenspach M, Brews JR, Mouret I, Schrimpf RD, Galloway KF (1994) Evaluation of SEGR threshold in power MOSFETs. IEEE Trans Nucl Sci 41:2160–2166

Ames B (2007) Intel tests chip design with 80-core processor. http://www.macworld.com/news/2007 /02/12/intel/

Amusan OA, Witulski AF, Massengill LW, Bhuva BL, Fleming PR, Alles ML, Sternberg AL, Black JD, Schrimpf RD (2006) Charge collection and charge sharing in a 130 nm CMOS technology. IEEE Trans Nucl Sci 53:3253–3258

Anderson T, Lee PA (1981) Fault tolerance: principles and practice. Prentice-Hall, Englewood Cliffs, NJ

Antola A, Erényi I, Scarabottolo N (1986) Transient fault management in systems based on the AMD 2900 microprocessors. Microprocess Microprogram 17:205–217

Applebaum SP (1965) Steady-state reliability of systems of mutually independent subsystems. IEEE Trans Reliab R-14:23–29

Arimoto K, Matsuda Y, Furutani K, Tsukude M, Ooishi T, Mashiko K, Fujishima K (1990) A speed-enhanced DRAM array architecture with embedded ECC. IEEE J Solid State Circuits 25:11–17

Armstrong DB (1966) On finding a nearly minimal set of fault detection tests for combinational logic nets. IEEE Trans Electron Comput EC-15:66–73

Asakura M, Matsuda Y, Hidaka H, Tanaka Y, Fujishima K (1990) An experimental 1-Mbit cache DRAM with ECC. IEEE J Solid State Circuits 25:5–10

V. Castano, I. Schagaev, *Resilient Computer System Design*,
DOI 10.1007/978-3-319-15069-7

Asanovic K, Bodik R, Catanzaro B, Gebis J, Husbands P, Keutzer K, Patterson D, Plishker W, Shalf J, Williams S, Yelick K (2006) The landscape of parallel computing research: a view from Berckeley. Technical report No. UCB/EECS-2006–183

Avizienis A (1971) Faulty-tolerant computing: an overview. Computer 4:5–8

Avizienis A (1976) Fault-tolerant systems. IEEE Trans Comput C-25:1304–1312

Avizienis A (1982) The four-universe information system model for the study of fault tolerance. In: Proceedings of the 12th annual international symposium on fault-tolerant computing, Santa Monica, CA, pp 6–13

Avizienis A, Kelly JPJ (1984) Fault tolerance by design diversity: concepts and experiments. Computer 17:67–80

Avizienis A, Laprie J-C, Randell B, Landwehr C (2004) Basic concepts and taxonomy of dependable and secure computing. IEEE Trans Dependable Secure Comput 1:11–33

Azadmanesh MH, Kieckhafer RM (2000) Exploiting omissive faults in synchronous approximate agreement. IEEE Trans Comput 49:1031–1042

Baeg S, Wen S, Wong R (2009) SRAM interleaving distance selection with a soft error failure model. IEEE Trans Nucl Sci 56:2111–2118

Barlow RE, Proschan F (1975) Statistical theory of reliability and life testing: probability models

Barth JL, LaBel KA, Poivey C (2004) Radiation assurance for the space environment. In: Integrated circuit design and technology, ICICDT'04. International conference on presented at the integrated circuit design and technology, pp. 323–333. doi:10.1109/ICICDT.2004.1309976

Baumann RC (2001) Soft errors in advanced semiconductor devices-part I: the three radiation sources. IEEE Trans Device Mater Reliab 1:17–22

Baumann R (2002) The impact of technology scaling on soft error rate performance and limits to the efficacy of error correction. In: International electron devices meeting (IEDM'02). Digest. pp 329–332

Baumann R (2005a) Soft errors in advanced computer systems. IEEE Des Test Comput 22:258–266

Baumann RC (2005b) Radiation-induced soft errors in advanced semiconductor technologies. IEEE Trans Dev Mater Rel 5:305–316

Baumann R, Hossain T, Murata S, Kitagawa H (1995) Boron compounds as a dominant source of alpha particles in semiconductor devices. In: Reliability physics symposium, 1995. 33rd annual proceedings, IEEE international, pp 297–302

Beaudry MD (1978) Performance-related reliability measures for computing systems. IEEE Trans Comput C-27:540–547

Becker HN, Miyahira TF, Johnston AH (2002) Latent damage in CMOS devices from single-event latchup. IEEE Trans Nucl Sci 49:3009–3015

Beitman BA (1988) n-channel MOSFET breakdown characteristics and modeling for p-well technologies. IEEE Trans Electron Dev 35:1935–1941

Bentoutou Y, Djaifri M (2008) Observations of single-event upsets and multiple-bit upsets in random access memories on-board the Algerian satellite. In: IEEE nuclear science symposium conference record (NSS'08), pp 2568–2570

Berger MJ, Coursey JS, Zucker MA, Chang J (2005) ESTAR, PSTAR, and ASTAR: computer programs for calculating stopping-power and range tables for electrons, protons, and helium ions (version 1.2.3). http://physics.nist.gov/Star. Accessed 11 Sept 2010.

Berka S (2009) Operating system documentation project. http://www.operating-system.org/betriebssys tem/_english/index.htm

Bernstein AV, Tomfield YL, Schagaev IV (1992) Storage unit with high reliability characteristics. I. Avtomat i Telemekh 145–152

Bernstein AV, Tomfield YL, Schagaev IV(1993) RAM of high reliability properties. II. Avtomat. i Telemekh 169–179

Binder D, Smith EC, Holman AB (1975) Satellite anomalies from galactic cosmic rays. IEEE Trans Nucl Sci 22:2675–2680. doi:10.1109/TNS.1975.4328188

Birolini A (2014) Reliability engineering, 7th edn. Springer, Heidelberg

Blake JB, Mandel R (1986) On-orbit observations of single event upset in Harris HM-6508 1K RAMS. IEEE Trans Nucl Sci 33:1616–1619. doi:10.1109/TNS.1986.4334651

Blandford JT, Waskiewicz AE, Pickel JC (1984) Cosmic ray induced permanent damage in MNOS EAROMs. IEEE Trans Nucl Sci 31:1568–1570

Bläser L (2006) A component language for structured parallel programming. Proc JMLC, Oxford, UK

Bläser L (2007) A high-performance operating system for structured concurrent programs. In: Proceedings of the workshop on Programming Languages and Operating Systems (PLOS), Oct 2007

Boatella C, Hubert G, Ecoffet R, Duzellier S (2009) ICARE on-board SAC-C: more than 8 years of SEU & MCU, analysis and prediction. IEEE, pp 369–374

Bodsberg L, Hokstad P (1995) A system approach to reliability and life-cycle cost of process safety-systems. IEEE Trans Reliab 44:179–186

Bogliolo A, Favalli M, Damiani M (2000) Enabling testability of fault-tolerant circuits by means of IDDQ checkable voters. IEEE Trans Very Large Scale Integr Syst 8:415–419

Bose RC, Ray-Chaudhuri DK (1960) On a class of error correcting binary group codes. Inf Control 3:68–79

Bougerol A, Miller F, Buard N (2008) SDRAM architecture & single event effects revealed with laser. In: 14th IEEE international on-line testing symposium (IOLTS'08). pp 283–288

Bougerol A, Miller F, Guibbaud N, Gaillard R, Moliere F, Buard N (2010) Use of laser to explain heavy ion induced SEFIs in SDRAMs. IEEE Trans Nucl Sci 57:272–278. doi:10.1109/TNS.2009.2037418

Bougerol A, Miller F, Guibbaud N, Leveugle R, Carriere T, Buard N (2011) Experimental demonstration of pattern influence on DRAM SEU and SEFI radiation sensitivities. IEEE Trans Nucl Sci 58:1032–1039

Bouricius WG, Carter WC, Schneider PR (1969) Reliability modeling techniques for self-repairing computer systems. In: Proceedings of the 1969 24th national conference, ACM' 69. ACM, New York, NY, pp 295–309

Bradley PD, Normand E (1998) Single event upsets in implantable cardioverter defibrillators. IEEE Trans Nucl Sci 45:2929–2940

Breuer MA (1973) Testing for intermittent faults in digital circuits. IEEE Trans Comput C-22:241–246

Brocklehurst S, Littlewood B, Olovsson T, Jonsson E (1994) On measurement of operational security. IEEE Aerosp Electron Syst Mag 9:7–16

Bryant R, O'Hallaron D (2002) Computer systems: a programmer's perspective. Prentice Hall, Upper Saddle River, NJ

Buchner S, Baze M, Brown D, McMorrow D, Melinger J (1997) Comparison of error rates in combinational and sequential logic. IEEE Trans Nucl Sci 44:2209–2216

Buckle R, Highleyman WH (2003) The new nonstop advanced architecture: a massive jump in processor reliability. Connection 24

Caldwell DW (1998) DSIGDE/PowerAnomaly Day300: analysis and resolution

Carter WC (1979) Hardware fault tolerance. In: Computing systems reliability. CUP Archive, pp 211–263

Carter WC, Bouricius WG (1971) A survey of fault tolerant computer architecture and its evaluation. Computer 4:9–16

Carter WC, Schneider PR (1968) Design of dynamically checked computers. In: IFIP congress (2). pp 878–883

Caywood JM, Prickett BL (1983) Radiation-induced soft errors and floating gate memories. In: 21st annual reliability physics symposium, 1983, pp 167–172

Cazeaux JM, Rossi D, Metra C (2004) New high speed CMOS self-checking voter. In: Proceedings of the international on-line testing symposium, 10th IEEE (IOLTS'04). IEEE Computer Society, Washington, DC, p 58

Cha H, Rudnick EM, Choi GS, Patel JH, Iyer RK (1993) A fast and accurate gate-level transient fault simulation environment. In: Digest of papers, presented at the twenty-third international symposium on the fault-tolerant computing, 1993. FTCS-23, pp 310–319

Chen CL, Hsiao MY (1984) Error-correcting codes for semiconductor memory applications: a state-of-the-art review. IBM J Res Dev 28:124–134

Claeys CL, Simoen E (2002) Radiation effects in advanced semiconductor materials and devices. Springer, Heidelberg

Coe T, Mathisen T, Moler C, Pratt V (1995) Computational aspects of the pentium affair. IEEE Comput Sci Eng 2:18–30

Conlon JC, Lilius WA, Tubbesing FH (1982) Test and evaluation of system reliability, availability, maintainability: a primer. Office of the Director, Defense Test and Evaluation, Under Secretary of Defense for Research and Engineering

Constantinescu C (2003) Trends and challenges in VLSI circuit reliability. IEEE Micro 23:14–19

Constantinescu C, Parulkar I, Harper R, Michalak S (2008) Silent data corruption – myth or reality? In: IEEE international conference on dependable systems and networks with FTCS and DCC, (DSN 2008), pp 108–109

Cottrell PE, Troutman RR, Ning TH (1979) Hot-electron emission in N-channel IGFET's. IEEE Trans Electron Dev 26:520–533

Crain SH, Velazco R, Alvarez MT, Bofill A, Yu P, Koga R (1999) Radiation effects in a fixed-point digital signal processor. Presented at the radiation effects data workshop, 1999, pp 30–34

Czajkowski DR, Pagey MP, Samudrala PK, Goksel M, Viehman MJ (2005) Low power, high-speed radiation hardened computer & flight experiment. Presented at the IEEE aerospace conference, 2005, pp 1–10

Czajkowski DR, Samudrala PK, Pagey MP (2006) SEU mitigation for reconfigurable FPGAs. Presented at the IEEE aerospace conference, 2006, p 7

DeAngelis D, Lauro JA (1976) Software recovery in the fault-tolerant spaceborne computer. , In: IEEE computer society sixth int. fault-tolerant computing symp, Pittsburg, PA, Digest. pp 143–148

Department of Electrical Engineering (1952) Whirlwind I master drawing list and general rack layout of computer. MIT, Cambridge, MA

Desko JC, Darwish MN, Dolly MC, Goodwin CA, Dawes WR, Titus JL (1990) Radiations hardening of a high voltage IC technology (BCDMOS). IEEE Trans Nucl Sci 37:2083–2088

Dhillon BS (2006) Maintainability, maintenance, and reliability for engineers. CRC, Boca Raton, FL

Dhillon YS, Diril AU, Chatterjee A, Metra C (2005) Load and logic co-optimization for design of soft-error resistant nanometer CMOS circuits. In: IEEE International on-line testing symposium, Los Alamitos, CA, pp 35–40

Dijkstra EW (1965) Solution of a problem in concurrent programming control. Commun ACM 8:569

Dixit A, Wood A (2011) The impact of new technology on soft error rates. In: IEEE international reliability physics symposium (IRPS), 2011, pp 5B.4.1–5B.4.7

Dodd PE (2005) Physics-based simulation of single-event effects. IEEE Trans Dev Mater Reliab 5:343–357

Dodd PE, Massengill LW (2003) Basic mechanisms and modeling of single-event upset in digital microelectronics. IEEE Trans Nucl Sci 50:583–602

Dodd PE, Shaneyfelt MR, Walsh DS, Schwank JR, Hash GL, Loemker RA, Draper BL, Winokur PS (2000) Single-event upset and snapback in silicon-on-insulator devices and integrated circuits. IEEE Trans Nucl Sci 47:2165–2174

Dreslinski RG, Wieckowski M, Blaauw D, Sylvester D, Mudge T (2010) Near-threshold computing: reclaiming Moore's law through energy efficient integrated circuits. Proc IEEE 98:253–266

Dufour C, Garnier P, Carriere T, Beaucour J, Ecoffet R, Labrunee M (1992) Heavy ion induced single hard errors on submicronic memories [for space application]. IEEE Trans Nucl Sci 39:1693–1697

Dugan JB, Trivedi KS (1989) Coverage modeling for dependability analysis of fault-tolerant systems. IEEE Trans Comput 38:775–787

Dunn M (1991) Designer fault models for VLSI. Presented at the IEE colloquium on design for testability, pp 4/1–4/5

Duzellier S, Falguere D, Ecoffet R (1993) Protons and heavy ions induced stuck bits on large capacity RAMs. In: Second European conference on radiation and its effects on components and systems (RADECS 93), pp 468–472

Duzellier S, Ecoffet R, Falguere D, Nuns T, Guibert L, Hajdas W, Calvert MC (1997) Low energy proton induced SEE in memories. IEEE Trans Nucl Sci 44:2306–2310

Dyer CS, Sims AJ, Farren J, Stephen J (1990) Measurements of solar flare enhancements to the single event upset environment in the upper atmosphere [avionics]. IEEE Trans Nucl Sci 37:1929–1937

Dyer CS, Truscott PR, Evans H, Sims AJ, Hammond N, Comber C (1996) Secondary radiation environments in heavy space vehicles and instruments. Adv Space Res 17:53–58

Eckert DI (2001) Odyssey MEEB analysis. Lockheed-Martin presentation

Ecoffet R, Duzellier S, Tastet P, Aicardi C, Labrunee M (1994) Observation of heavy ion induced transients in linear circuits. In: IEEE radiation effects data workshop, 1994, pp 72–77

ECSS (2007) Space engineering: methods for the calculation of radiation received and its effects, and a policy for design margins—ECSS-E-10–12 Draft 0.5

Edwards R, Dyer C, Normand E (2004) Technical standard for atmospheric radiation single event effects, (SEE) on avionics electronics. Presented at the IEEE radiation effects data workshop, 2004, pp 1–5

EIA/JEDEC Standard (1996) Test procedures for the measurement of single-event effects in semiconductor devices from heavy ion irradiation. EIA/JEDEC Standard

Elsayed EA (1996) Reliability engineering. Prentice Hall, Upper Saddle River, NJ, Har/Dsk. edition

Everett R, Swain F (1947) Report R-127, Whirlwind I computer block diagrams. MIT Servomechanisms Laboratory

Felix JA, Shaneyfelt MR, Schwank JR, Dalton SM, Dodd PE, Witcher JB (2007) Enhanced degradation in power MOSFET devices due to heavy ion irradiation. IEEE Trans Nucl Sci 54:2181–2189

Fieseler PD, Ardalan SM, Frederickson AR (2002) The radiation effects on Galileo spacecraft systems at Jupiter. IEEE Trans Nucl Sci 49:2739–2758. doi:10.1109/TNS.2002.805386

Fischer TA (1987) Heavy-ion-induced, gate-rupture in power MOSFETs. IEEE Trans Nucl Sci 34:1786–1791

Fisher JA (1983) Very long instruction word architectures and the ELI-512. Proceedings of the 10th annual international symposium on computer architecture. ACM, Stockholm, Sweden, pp 140–150

Fleetwood D, Pantelides S, Schrimpf R (2008) Oxide traps, border traps, and interface traps in SiO2. In: Fleetwood D, Pantelides S, Schrimpf R (eds) Defects in microelectronic materials and devices. CRC, Boca Raton, FL

Flynn M (1972) Some computer organizations and their effectiveness. IEEE Trans Comput C-21:948

Fortes JAB, Raghavendra CS (1985) Gracefully degradable processor arrays. IEEE Trans Comput C-34:1033–1044

Franklin M, Saluja KK (1995) Embedded RAM testing. Records of the 1995 I.E. international workshop on memory technology, design and testing, pp 29–33

Franzon P, Harrod W, Hill K, Hiller J, Karp S, Keckler S, Klein D, Lucas R (2010) Guide for HPC applications on IBM power 755 system

Friedman AD (1967) Fault detection in redundant circuits. IEEE Trans Electr Comput EC-16:99–100

Furutani K, Arimoto K, Miyamoto H, Kobayashi T, Yasuda K, Mashiko K (1989) A built-in Hamming code ECC circuit for DRAMs. IEEE J Solid State Circuits 24:50–56

Galey JM, Norby RE, Roth JP (1961) Techniques for the diagnosis of switching circuit failures. In: Proceedings of the second annual symposium on switching circuit theory and logical design (SWCT 1961), pp 152–160

Gaisler J (2002) A portable and fault-tolerant microprocessor based on the SPARC v8 architecture. European Space Agency, Noorwijk. Conference on dependable systems and networks, DSN 2002. doi:10.1109/DSN.2002.1028926

Geer D (2007) For programmers, multicore chips mean multiple challenges. Computer 40:17–19

Geppert L (2004) A static RAM says goodbye to data errors [radiation induced soft errors]. IEEE Spectr 41:16–17

Gnedenko B, Pavlov IV, Ushakov IA (1999) Statistical reliability engineering, 1st edn. Wiley-Interscience, Hoboken NJ

Goldstein L (1979) Controllability/observability analysis of digital circuits. IEEE Trans Circuit Syst 26:685–693

Goldstine A, Goldstine HH (1946) ENIAC, the electronic numerical integrator. Math Tables Other Aids Comput 1:97–110

Gosden J (1966) Explicit parallel processing description and control in programs for multi- and uni- processor computers. In: Proc of AFIPS'66, 7–10 Nov 1966. ACM, New York, pp 651–660

Gössel M, Ocheretny V, Sogomonyan E, Marienfeld D (2008) New methods of concurrent checking. Springer, The Netherlands

Goth G (2009) Entering a parallel universe. Commun ACM 52:15

Gountanis RJ, Viss NL (1966) A method of processor selection for interrupt handling in a multiprocessor system. Proc IEEE 54:1812–1819

Gregory BL, Shafer BD (1973) Latch-up in CMOS integrated circuits. IEEE Trans Nucl Sci 20:293–299

Guenzer CS, Wolicki EA, Allas RG (1979) Single event upset of dynamic rams by neutrons and protons. IEEE Trans Nucl Sci 26:5048–5052

Gutknecht J (2006) The dining philosophers problem revisited, JMLC 2006. Lect Notes Comput Sci 4228:377–382

Hamming RW (1950) Error correction and error detection coding. Bell Syst Tech J 29:147–160

Hana HH, Johnson BW (1986) Concurrent error detection in VLSI circuits using time redundancy. In: Proc IEEE Southeastcon 1986 regional conf., pp 208–212

Haraszti TP (2000) CMOS memory circuits. Springer, New York, NY

Harboe-Sorensen R, Guerre F-X, Lewis G (2007) Heavy-ion SEE test concept and results for DDR-II memories. IEEE Trans Nucl Sci 54:2125–2130. doi:10.1109/TNS.2007.909747

Hauge PS, Ziegler JF, Srinivasan GR (1996) Special issue: terrestrial cosmic rays and soft errors. IBM J Res Dev. http://portal.acm.org/citation.cfm?id=226354. Accessed 7 Jan 2010

Hawkins C, Keshavarzi A, Segura J (2003) CMOS IC nanometer technology failure mechanisms. The custom integrated circuits conference, 2003. Proceedings of the IEEE, pp 605–611

Hayes JP (1975) Detection of pattern-sensitive faults in random-access memories. IEEE Trans Comput C-24:150–157

Hazucha P, Svensson C (2000) Impact of CMOS technology scaling on the atmospheric neutron soft error rate. IEEE Trans Nucl Sci 47:2586–2594

Hazucha P, Karnik T, Maiz J, Walstra S, Bloechel B, Tschanz J, Dermer G, Hareland S, Armstrong P, Borkar S (2003) Neutron soft error rate measurements in a 90-nm CMOS process and scaling trends in SRAM from 0.25- mu;m to 90-nm generation. Technical digest, IEEE international electron devices meeting. (IEDM'03). pp 21.5.1–21.5.4

Heidel DF, Marshall PW, LaBel KA, Schwank JR, Rodbell KP, Hakey MC, Berg MD, Dodd PE, Friendlich MR, Phan AD, Seidleck CM, Shaneyfelt MR, Xapsos MA (2008) Low energy proton single-event-upset test results on 65 nm SOI SRAM. IEEE Trans Nucl Sci 55:3394–3400

Heise B (2009) Computerproblemelegen check-in-system der lufthansalahr. http://www.heise.de/newsticker/meldung/Computerprobleme-legen-Check-in-System-der-Lufthansa-lahm-798193.html

Hennessy JL, Patterson DA (2006) Computer architecture: a quantitative approach, 4th edn. Morgan Kaufmann, Burlington, MA

Hentschke R, Marques F, Lima F, Carro L, Susin A, Reis R (2002) Analyzing area and performance penalty of protecting different digital modules with hamming code and triple modular redundancy. In: Proceedings of the 15th symposium on integrated circuits and systems design, SBCCI'02. IEEE computer society, Washington, DC, p 95

Hill MD, Rajwar R (2001) The rise and fall of multiprocessor papers in ISCA [www.document]. The rise and fall of multiprocessor papers in the international symposium on computer architecture (ISCA). http://pages.cs.wisc.edu/~markhill/mp2001.html. Accessed 4 Aug 2010

Hofstra University (1999) History in the computing curriculum. Appendix 4 1950–59. www.comphist.org/pdfs/CompHist_9812tla4.pdf

Hohl JH, Galloway KF (1987) Analytical model for single event burnout of power MOSFETs. IEEE Trans Nucl Sci 34:1275–1280

Holland JH (1960) Iterative circuit computers. In: Western joint IRE- AIEE-ACM computer conference, May 3–5, 1960, San Francisco, CA, pp. 259–265

Howe CL, Weller RA, Reed RA, Mendenhall MH, Schrimpf RD, Warren KM, Ball DR, Massengill LW, LaBel KA, Howard JW, Haddad NF (2005) Role of heavy-ion nuclear reactions in determining on-orbit single event error rates. IEEE Trans Nucl Sci 52:2182–2188

Hsiao MY (1970) A class of optimal minimum odd-weight-column SEC-DED codes. IBM J Res Dev 14:395–401

Hsieh CM, Murley PC, O'Brien RR (1981) A field-funneling effect on the collection of alpha-particle-generated carriers in silicon devices. IEEE Electron Device Lett 2:103–105

Hsieh C-M, Murley PC, O'Brien RR (1983) Collection of charge from alpha-particle tracks in silicon devices. IEEE Trans Electron Dev 30:686–693

Hughes HL, Benedetto JM (2003) Radiation effects and hardening of MOS technology: devices and circuits. IEEE Trans Nucl Sci 50:500–521

Hugue MM, Purtilo J (2002) Guerrilla tactics: motivating design patterns for high-dependability applications. In: Proceedings of the 27th annual NASA Goddard/IEEE software engineering workshop, pp 33–39

Hutcheson GD (2009) The economic implications of Moore's Law. In: Into the nano era. pp 11–38

Ibe E, Chung SS, Wen S, Yamaguchi H, Yahagi Y, Kameyama H, Yamamoto S, Akioka T (2006) Spreading diversity in multi-cell neutron-induced upsets with device scaling. Presented at the IEEE custom integrated circuits conference, 2006, pp 437–444

Ibe E, Taniguchi H, Yahagi Y, K S, Toba T (2010) Impact of scaling on neutron-induced soft error in SRAMs from a 250 nm to a 22 nm design rule. IEEE Trans Electron Dev 57:1527–1538

IBM (2007) IBM Power6 microprocessor and IBM System p 570

Irom F, Nguyen DN (2007) Single event effect characterization of high density commercial NAND and NOR nonvolatile flash memories. IEEE Trans Nucl Sci 54:2547–2553

ITRS (2011) International technology roadmap for semiconductors

JEDEC "JESD89-3A," (2007) www.jedec.org/sites/default/files/docs/JESD89-3A.pdf

Jennings BF (1990) Fault detection in microprocessor based systems. In: IEE colloquium on fault tolerant techniques, pp 7/1–718

Johansson K, Dyreklev P, Granbom O, Calver MC, Fourtine S, Feuillatre O (1998) In-flight and ground testing of single event upset sensitivity in static RAMs. IEEE Trans Nucl Sci 45:1628–1632. doi:10.1109/23.685251

Johnson BW (1989) The design and analysis of fault tolerant digital systems. Addison-Wesley, Reading, MA

Johnson BW, Aylor JH, Hana HH (1988) Efficient use of time and hardware redundancy for concurrent error detection in a 32-bit VLSI adder. IEEE J Solid State Circuits 23:208–215

Johnson GH, Schrimpf RD, Galloway KF, Koga R (1992) Temperature dependence of single-event burnout in n-channel power MOSFETs [for space application]. IEEE Trans Nucl Sci 39:1605–1612

Johnston AH (1996) The influence of VLSI technology evolution on radiation-induced latchup in space systems. IEEE Trans Nucl Sci 43:505–521

Johnston AH, Hughlock BW, Baze MP, Plaag RE (1991) The effect of temperature on single-particle latchup. IEEE Trans Nucl Sci 38:1435–1441

Kaczer B, Arkhipov V, Degraeve R et al (2005) Temperature dependence of the negative bias temperature instability in the framework of dispersive transport. Appl Phys Lett 86:143506

Kadayif I, Sen H, Koyuncu S (2010) Modeling soft errors for data caches and alleviating their effects on data reliability. Microprocess Microsyst 34:200–214

Kaegi T, Schagaev I (2013) System software support of hardware deficiency. In: ITACS 2013, ISBN 978- 0-9575049-0-5

Kaegi-Trachsel T et al (2009) Hardware testing on the level of tasks. Preprints of the 30th IFAC workshop on real-time programming and 4th international workshop on real-time software, pp 79–84. https://fedcsis.org/2009/

Karimi F, Lombardi F (2002) A scan-BIST environment for testing embedded memories. In: Proceedings of the eighth IEEE international on-line testing workshop. pp 211–217

Karnik T, Hazucha P (2004) Characterization of soft errors caused by single event upsets in CMOS processes. IEEE Trans Depend Secure Comput 1:128–143

Kato M, Watanabe K, Okabe T (1989) Radiation effects on ion-implanted silicon-dioxide films. IEEE Trans Nucl Sci 36:2199–2204

Katz R, Barto R, McKerracher P, Carkhuff B, Koga R (1994) SEU hardening of field programmable gate arrays (FPGAs) for space applications and device characterization. IEEE Trans Nucl Sci 41:2179–2186

Katz R, LaBel K, Wang JJ, Cronquist B, Koga R, Penzin S, Swift G (1997) Radiation effects on current field programmable technologies. IEEE Trans Nucl Sci 44:1945–1956

Kaufman L, Johnson BW (2001) Embedded digital system reliability and safety analysis. NUREG/GR-0020

Kim J, Hardavellas N, Mai K, Falsafi B, Hoe J (2007) Multi-bit error tolerant caches using two-dimensional error coding. In: Proceedings of the 40th annual IEEE/ACM international symposium on microarchitecture (MICRO 40), pp 197–209

Kinnersley B (2009) The language list. http://people.ku.edu/~nkinners/LangList/Extras/langlist.htm

Kish LB (2002) End of Moore's law: thermal (noise) death of integration in micro and nano electronics. Phys Lett A 305:144–149

Koga R, Kolasinski WA (1989) Heavy ion induced snapback in CMOS devices. IEEE Trans Nucl Sci 36:2367–2374

Koga R, Kolasinski WA, Marra MT, Hanna WA (1985) Techniques of microprocessor testing and SEU-rate prediction. IEEE Trans Nucl Sci 32:4219–4224

Koga R, Crain WR, Crawford KB, Lau DD, Pinkerton SD, Yi BK, Chitty R (1991) On the suitability of non-hardened high density SRAMs for space applications. IEEE Trans Nucl Sci 38:1507–1513

Koga R, Crawford KB, Grant PB, Kolasinski WA, Leung DL, Lie TJ, Mayer DC, Pinkerton SD, Tsubota TK (1993a) Single ion induced multiple-bit upset in IDT 256K SRAMs. In: Second European conference on radiation and its effects on components and systems (RADECS 93), pp 485–489

Koga R, Pinkerton SD, Lie TJ, Crawford KB (1993b) Single-word multiple-bit upsets in static random access devices. IEEE Trans Nucl Sci 40:1941–1946

Koga R, Crain SH, Yu P, Crawford KB (2001a) SEE sensitivity determination of high-density DRAMs with limited-range heavy ions. In: IEEE radiation effects data workshop, 2001, pp 182–189

Koga R, Yu P, Crawford KB, Crain SH, Tran VT (2001b) Permanent single event functional interrupts (SEFIs) in 128- and 256-megabit synchronous dynamic random access memories (SDRAMs). In: IEEE radiation effects data workshop, pp 6–13

Kogge P, Bergman K, Borkar S, Campbell D, Carlson W, Dally W, Denneau M, Franzon P, Harrod W, Hill K, Hiller J, Karp S, Keckler S, Klein D, Lucas R, Richards M, Scarpelli A, Scott S, Snavely A, Sterling T, Williams RS, Yelick KA (2008) Exascale computing study: technology challenges in achieving exascale systems. DARPA IPTO, technical report DARPA-2008-13, Sept 2008

Koren I, Koren Z (1998) Defect tolerance in VLSI circuits: techniques and yield analysis. Proc IEEE 86:1819–1838

Koren I, Krishna CM (2007) Fault-tolerant systems. Morgan Kaufmann, San Francisco, CA

Koren I, Singh AD (1990) Fault tolerance in VLSI circuits. Computer 23:73–83

Kovalenko IN, Kuznetzov NY, Pegg PA (1997) Mathematical theory of reliability of time dependent systems with practical applications, 1st edn. Wiley, New York, NY

Kulkarni GV, Nicola FV, Trivedi SK (1987) Effects of checkpointing and queueing on program performance. Duke University, Durham, NC, USA

Kumar S, Kim C, Sapatnekar S (2006) An analytical model for negative bias temperature instability. In: Proceedings of IEEE/ACM ICCAD, pp 493–496

LaBel K, Stassinopoulos EG, Brucker GJ, Stauffer CA (1992) SEU tests of a 80386 based flight-computer/data-handling system and of discrete PROM and EEPROM devices, and SEL tests of discrete 80386, 80387, PROM, EEPROM and ASICs, Workshop record. IEEE radiation effects data workshop, 1992, pp 1–11

LaBel KA, Gates MM, Moran AK, Kim HS, Seidleck CM, Marshall P, Kinnison J, Carkhuff B (1996) Radiation effect characterization and test methods of single-chip and multi-chip stacked 16 Mbit DRAMs. IEEE Trans Nucl Sci 43:2974–2981

Ladbury R, Berg MD, Kim H, LaBel K, Friendlich M, Koga R, George J, Crain S, Yu P, Reed RA (2006) Radiation performance of 1 Gbit DDR SDRAMs fabricated in the 90 nm CMOS technology node. In: IEEE radiation effects data workshop, pp 126–130

Lala JH, Harper RE (1994) Architectural principles for safety-critical real-time applications. Proc IEEE 82:25–40

Lamport L (1983) The weak Byzantine Generals problem. J ACM 30:668–676

Lamport L, Melliar-Smith P (1985) Synchronising clocks in the presence of faults. J ACM 32 (1):52–78

Landis DL (1989) A self-test system architecture for reconfigurable WSI. In: International test conference proceedings. Meeting the tests of time, pp 275–282

Landwehr CE, Bull AR, McDermott JP, Choi WS (1994) A taxonomy of computer program security flaws. ACM Comput Surv 26:211–254

LaPedus M, EETimes (2003) Intel gears up 90-nm processor, chip set rollout. http://www.eetimes.com/conf/idf/showArticle.jhtml?articleID=10800811&kc=3172

Laprie J (1995) Dependability – its attributes, impairments and means. In: Randell B, Laprie J, Kopetz H, Littlewood B (eds) Predictably dependable computing systems. Springer, Heidelberg, pp 1–24

Laprie J-C (2008) From dependability to resilience 8, G8–G9

Laprie J, Avizienis A (1986) Dependable computing: from concepts to design diversity. In: Proc IEEE, pp 629–638

Laprie JCC, Avizienis A, Kopetz H (eds) (1992) Dependability: basic concepts and terminology. Springer, New York, NY

Latchoumy P, Sheik P, Khader A (2011) Survey on fault tolerance in grid computing

Lawrence RK (2007) Radiation characterization of 512Mb SDRAMs. In: IEEE radiation effects data workshop, pp 204–207

Leavy JF, Poll RA (1969) Radiation-induced integrated circuit latchup. IEEE Trans Nucl Sci 16:96–103

Li J, Swartzlander E (1992) Concurrent error detection in ALUs by recomputing with rotated operands. In: Proceedings of the IEEE international workshop on defect and fault tolerance in VLSI systems, pp 109–116

Lie CH, Hwang CL, Tillman FA (1977) Availability of maintained systems: a state-of-the-art survey. AIIE Transact 9:247–259

Lin S, Costello DJ (1983) Error control coding: fundamentals and applications. Prentice Hall, Upper Saddle River, NJ

Mahout G, Pearce M, Andrieux M-L, Arvidsson C-B, Charlton DG, Dinkespiler B, Dowell JD, Gallin-Martel L, Homer RJ, Jovanovic P, Kenyon IR, Kuyt G, Lundquist J, Mandic I, Martin O, Shaylor HR, Stroynowski R, Troska J, Wastie RL, Weidberg AR, Wilson JA, Ye J (2000) Irradiation studies of multimode optical fibres for use in ATLAS front-end links. Nucl Instrum Meth A 446:426–434

Maiz J, Hareland S, Zhang K, Armstrong P (2003) Characterization of multi-bit soft error events in advanced SRAMs. In: IEEE international electron devices meeting (IEDM'03), Technical digest, pp 21.4.1–21.4.4

Mao W, Gulati RK, Goel DK, Ciletti MD (1990) QUIETEST: a quiescent current testing methodology for detecting leakage faults. In: IEEE international conference on computer-aided design (ICCAD-90), Digest of technical papers, pp 280–283

Marshall RW (1963) Microelectronic devices for application in transient nuclear radiation environments. In: IEEE transactions on aerospace and navigational electronics, technical paper, pp 1.4.1–1–1.4.1–6

Mathew B, Saab DG (1993) Partial reset: an inexpensive design for testability approach. In: Proceedings of the [4th] European conference on design automation, with the European event in ASIC design. pp 151–155

Matijevic J (1996) Mars Pathfinder microrover – implementing a low cost planetary mission experiment. John Hopkins Applied Physics Laboratory, pp 16–19

Matsunaga J, Momose H, Iizuka H, Kohyama S (1980) Characterization of two step impact ionization and its influence in NMOS and PMOS VLSI's. Presented at the international electron devices meeting. pp 736–739

Mavis DG (2002) Single event transient phenomena—Challenges and solutions. In: Microelectronics reliability and qualification workshop

Mavis DG, Eaton PH (2002) Soft error rate mitigation techniques for modern microcircuits. In: 40th annual reliability physics symposium proceedings, pp 216–225

May T (1979) Soft errors in VLSI: present and future. IEEE Trans Component Hybrid Manufact Technol 2:377–387

May TC, Woods MH (1979) Alpha-particle-induced soft errors in dynamic memories. IEEE Trans Electron Dev 26:2–9

May TC, Scott GL, Meieran ES, Winer P, Rao VR (1984) Dynamic fault imaging of VLSI random logic devices. In: 22nd Annual reliability physics symposium, pp 281–283

McCluskey EJ, Tseng C-W (2000) Stuck-fault tests vs. actual defects. In: Proceedings of the international test conference, 2000, pp 336–342

McEvoy D (1981) The architecture of Tandem's NonStop system. In: Proceedings of the ACM'81 Conference, p 245

McMahan MA, Leitner D, Gimpel T, Morel J, Ninemire B, Siero R, Silver C, Thatcher R (2004) A 16 MeV/nucleon cocktail for heavy ion testing. In: IEEE radiation effects data workshop, 2004, pp 156–159

Mei KCY (1974) Bridging and stuck-at faults. IEEE Trans Comput C-23:720–727

Messenger GC, Ash MS (1992) The effects of radiation on electronic systems, 2nd edn. Springer, New York, NY

Metra C, Favalli M, Ricco B (1997) Compact and low power on-line self-testing voting scheme. Proceedings of the IEEE international symposium on defect and fault tolerance in VLSI systems. pp 137–145

Meyer FJ, Pradhan DK (1991) Consensus with dual failure modes. IEEE Trans Parallel Distr Syst 2:214–222

Miller LS, Mullin JB (1991) Electronic materials: from silicon to organics. Springer, New york, NY

Miller LA, Brice DK, Prinja AK, Picraux ST (1994) Molecular dynamics simulations of bulk displacement threshold energies in Si. Radiation effects and defects in solids: incorporating plasma science and plasma technology 129, 127

Monkman S, Schagaev I (2013) Redundancy + repeatability = recoverability. Electronics 2:212–233. doi:10.3390/electronics2030212, ISSN 2079-9292

Moon TK (2005) Error correction coding: mathematical methods and algorithms. Wiley, Hoboken NJ

Mrstik BJ, Hughes HL, McMarr PJ, Lawrence RK, Ma DI, Isaacson IP, Walker RA (2000) Hole and electron trapping in ion implanted thermal oxides and SIMOX. IEEE Trans Nucl Sci 47:2189–2195

Mukherjee SS, Emer J, Fossum T, Reinhardt SK (2004) Cache Scrubbing in Microprocessors: Myth or Necessity? Proceedings of the 10th IEEE Pacific rim international symposium on dependable computing (PRDC'04). IEEE Computer Society, Washington, DC, pp 37–42

Nair R (2002) Effect of increasing chip density on the evolution of computer architectures. IBM J Res Dev 46:223–234

Nair R, Thatte SM, Abraham JA (1978) Efficient algorithms for testing semiconductor random-access memories. IEEE Trans Comput C-27:572–576

Naseer R, Bhatti RZ, Draper J (2006) Analysis of soft error mitigation techniques for register files in IBM Cu-08 90nm technology. In: 49th IEEE international Midwest symposium on circuits and systems (MWSCAS'06), pp 515–519

NASNGSFC Landsat-7 Project Office. Private communication, 1995

Neuberger G, de Lima F, Carro L, Reis R (2003) A multiple bit upset tolerant SRAM memory. ACM Trans Des Autom Electron Syst 8:577–590

Neuberger G, de Lima Kastensmidt FG, Reis R (2005) An automatic technique for optimizing Reed-Solomon codes to improve fault tolerance in memories. IEEE Design Test Comput 22:50–58

Newell A (1960) A on programming a highly parallel machine to be an intelligent technician, IRE-AIEE-ACM '60 (Western), May 3–5. ACM, New York, NY, pp 267–282

Nguyen DN, Guertin SM, Swift GM, Johnston AH (1999) Radiation effects on advanced flash memories. IEEE Trans Nucl Sci 46:1744–1750

Nicolaidis M (ed) (2010) Soft errors in modern electronic systems, 1st edn. Springer, New York, NY

Ning TH, Yu HN (1974) Optically induced injection of hot electrons into SiO2. J Appl Phys 45:5373–5378

Nishioka Y, Ohyu K, Ohji Y, Kato M, da Silva EF, Ma TP (1989) Radiation hardened micron and submicron MOSFETs containing fluorinated oxides. IEEE Trans Nucl Sci 36:2116–2123

Normand E, Wert JL, Quinn H, Fairbanks TD, Michalak S, Grider G, Iwanchuk P, Morrison J, Wender S, Johnson S (2010) First record of single-event upset on ground, cray-1 computer at Los Alamos in 1976. IEEE Trans Nucl Sci 57:3114–3120

Northcliffe L, Schilling R (1970) Range and stopping-power tables for heavy ions. At Data Nucl Data Tables 7:233–463

Oh N, Shirvani PP, McCluskey EJ (2002a) Control-flow checking by software signatures. IEEE Trans Reliab 51:111–122

Oh N, Shirvani PP, McCluskey EJ (2002b) Error detection by duplicated instructions in super-scalar processors. IEEE Trans Reliab 51:63–75

Oldham TR, Ladbury RL, Friendlich M, Kim HS, Berg MD, Irwin TL, Seidleck C, LaBel KA (2006) SEE and TID characterization of an advanced commercial 2Gbit NAND flash nonvolatile memory. IEEE Trans Nucl Sci 53:3217–3222

Oldham TR, Suhail M, Friendlich MR, Carts MA, Ladbury RL, Kim HS, Berg MD, Poivey C, Buchner SP, Sanders AB, Seidleck CM, LaBel KA (2008) TID and SEE response of advanced 4G NAND flash memories. In: IEEE radiation effects data workshop, pp 31–37

Olsen J, Becher PE, Fynbo PB, Raaby P, Schultz J (1993) Neutron-induced single event upsets in static RAMS observed a 10 km flight attitude. IEEE Trans Nucl Sci 40:74–77

Owens BD, Adams ME, Benzce WJ, Green G, Shestople P (2006) The effects of radiation events on gravity probe B. IEEE Trans Nucl Sci (in press)

Pankratius V, Jannesari A, Tichy WF (2009) Parallelizing Bzip2: a case study in multicore software engineering. IEEE Softw 26:70–77

Patel JH, Fung LY (1982) Concurrent error detection in ALU's by recomputing with shifted operands. IEEE Trans Comput C-31:589–595

Paul S, Cai F, Zhang X, Bhunia S (2011) Reliability-driven ECC allocation for multiple bit error resilience in processor cache. IEEE Trans Comput 60:20–34

Pellish JA, Reed RA, Schrimpf RD, Alles ML, Varadharajaperumal M, Niu G, Sutton AK, Diestelhorst RM, Espinel G, Krithivasan R, Comeau JP, Cressler JD, Vizkelethy G, Marshall PW, Weller RA, Mendenhall MH, Montes EJ (2006) Substrate engineering concepts to mitigate charge collection in deep trench isolation technologies. IEEE Trans Nucl Sci 53:3298–3305

Pickel JC, Blandford JT (1978) Cosmic ray induced in MOS memory cells. IEEE Trans Nucl Sci 25:1166–1171

Pickel JC, Blandford JT (1980) Cosmic-ray-induced errors in MOS devices. IEEE Trans Nucl Sci 27:1006–1015

Pierce WH (1965) Failure-tolerant computer design. Academic, New York, NY

Plyaskota S, Schagaev I (1995) Economic effectiveness of fault tolerance. Automation and Remote Control 07:1017–1026

Podgorsak EB (2009) Radiation physics for medical physicists. Springer, New York, NY

Pomeranz I, Reddy SM (1993) Classification of faults in synchronous sequential circuits. IEEE Trans Comput 42:1066–1077

Pouponnot ALR (2005) Strategic use of SEE mitigation techniques for the development of the ESA microprocessors: past, present and future. In: Proceedings of the 11th IEEE international on-line testing symposium, (IOLTS'05). IEEE Computer Society, Washington, DC, pp 319–323

Power 6 Specs: IBM Power6 Microprocessor and IBM System p 570, 2007

Prasad AVSS, Agrawal VD, Atre MV (2002) A new algorithm for global fault collapsing into equivalence and dominance sets. In: Proceedings of the international test conference, 2002, pp 391–397

Price D (1995) Pentium FDIV flaw-lessons learned. IEEE Micro 15:86–88

Pritchard BE, Swift GM, Johnston AH (2002) Radiation effects predicted, observed, and Galileo compared for spacecraft systems. In: 2002 IEEE radiation effects data workshop. California Institute of Technology, Pasadena, CA, USA

Puchner H, Kapre R, Sharifzadeh S, Majjiga J, Chao R, Radaelli D, Wong S (2006) Elimination of single event latchup in 90 nm SRAM technologies. Presented at the 44th annual reliability physics symposium proceedings. IEEE International, pp 721–722

Qian Y (2008) Information assurance: dependability and security in networked systems. Morgan Kaufmann, Burlington, MA

Quinn H, Graham P, Krone J, Caffrey M, Rezgui S (2005) Radiation-induced multi-bit upsets in SRAM-based FPGAs. IEEE Trans Nucl Sci 52:2455–2461

Ramanarayanan R, Degalahal VS, Krishnan R, Kim J, Narayanan V, Xie Y, Irwin MJ, Unlu K (2009) Modeling soft errors at the device and logic levels for combinational circuits. IEEE Trans Depend Secure Comput 6:202–216

Ravishankar K, Iyer ZK (2003) Hardware and software error detection

Reed IS, Solomon G (1960) Polynomial codes over certain finite fields. J Soc Ind Appl Math 8:300–304

Reed RA, Carts MA, Marshall PW, Marshall CJ, Buchner S, La Macchia M, Mathes B, McMorrow D (1996) Single Event Upset cross sections at various data rates. IEEE Trans Nucl Sci 43:2862–2867

Reed RA, Carts MA, Marshall PW, Marshall CJ, Musseau O, McNulty PJ, Roth DR, Buchner S, Melinger J, Corbiere T (1997) Heavy ion and proton-induced single event multiple upset. IEEE Trans Nucl Sci 44:2224–2229

Reed RA, Weller RA, Schrimpf RD, Mendenhall MH, Warren KM, Massengill LW (2006) Implications of nuclear reactions for single event effects test methods and analysis. IEEE Trans Nucl Sci 53:3356–3362. doi:10.1109/TNS.2006.885950

Reviriego P, Maestro JA, Cervantes C (2007) Reliability Analysis of Memories Suffering Multiple Bit Upsets. IEEE Trans Device Mater Reliab 7:592–601

Reviriego P, Maestro JA, Baeg S, Wen S, Wong R (2010) Protection of memories suffering MCUs through the selection of the optimal interleaving distance. IEEE Trans Nucl Sci 57:2124–2128

Reynolds DA, Metze G (1978) Fault detection capabilities of alternating logic. IEEE Trans Comput C-27:1093–1098

Roche P, Gasiot G (2005) Impacts of front-end and middle-end process modifications on terrestrial soft error rate. IEEE Trans Device Mater Reliab 5:382–396

Rockett LR (1988) An SEU-hardened CMOS data latch design. IEEE Trans Nucl Sci 35:1682–1687

Roy RK, Niermann TM, Patel JH, Abraham JA, Saleh RA (1988) Compaction of ATPG-generated test sequences for sequential circuits. In: IEEE international conference on computer-aided design (ICCAD-88). Digest of technical papers, pp 382–385

Sager D, Group DP, Corp I (2001) The microarchitecture of the pentium 4 processor. Intel Technol J 1:2001

Saleh AM, Serrano JJ, Patel JH (1990) Reliability of scrubbing recovery-techniques for memory systems. IEEE Trans Reliab 39:114–122

Sandireddy RKKR, Agrawal VD (2005) Diagnostic and detection fault collapsing for multiple output circuits. In: Proceedings of the design, automation and test in Europe, vol 2, pp 1014–1019

Schagaev I (1986a) Detecting the defective computer in two-unit, fault tolerant system having a sliding stand- by units. Automatic and Remote Control 5:143–150

Schagaev I (1986b) Algorithms of computation recovery, Automatic and remote control 7. Plenum, New York, NY

Schagaev I (1987) Algorithms to restoring a computing process, Automatic and Remote Control 7. Plenum, New York, NY

Schagaev I (1989) Yet another approach to classification of redundancy. In: Proceedings of FTSD. Prague, Czeschoslovakia, pp 485–490

Schagaev I (1990) Yet another approach to classification of redundancy. In: Proceedings of the seventh IMECO symposium technical diagnostics, 17–19 Sept, Helsinki, pp 485–492

Schagaev JZI (2001) Redundancy classification and its applications for fault tolerant computer design. IEEE TESADI-01

Schagaev I (2008) Reliability of malfunction tolerance. In: International multiconference on computer science and information technology (IMCSIT), Wisla, Poland, pp. 733–737

Schagaev I (2009) ERA: embedded reconfigurable architecture – past present and future

Schagaev I, Buhanova G (2001) Comparative study of fault tolerant RAM structures, in: IEEE DSN01. Presented at the IEEE DSN01, Goteborg

Schagaev I, Kaegi T, Gutknetch J (2010) ERA: evolving reconfigurable architecture. In: Proceedings of 11th ACIS, presented at the international conference on software engineering artificial intelligence, Networking and Parallel/Distributed Computing, London

Schagaev I, Bacon E, Hagel G, Charnine M, Kirk B, Kravtsov G (2013) Weduca: Web-enhanced design of university curricula, 07/2013. In: Proceeding of: FECS'13. ISBN: 1-60132-235-6

Schagaev I et al (2014) http://worldcomp-proceedings.com/proc/p2014/FCS3102.pdf

Schindlbeck G (2005) Types of soft errors in DRAMs. In: 8th European conference on radiation and its effects on components and systems (RADECS 2005), pp PE1–1–PE1–5

Schroder DK, Babcock JA (2003) Negative bias temperature instability: Road to cross in deep submicron silicon semiconductor manufacturing. J Appl Phys 94:1–18

Schwank JR, Dodd PE (2003) Charge collection in SOI capacitors and circuits and its effect on SEU hardness. IEEE Trans Nucl Sci 2937–2947

Schwartz E (1961) An automatic sequencing procedure with application to parallel programming. JACM 8(4):513–537

Segura J, Hawkins CF (2005) Bridging defects, CMOS electronics. Wiley, New York, NY, pp 199–222

Seifert N, Zhu X, Massengill LW (2002) Impact of scaling on soft-error rates in commercial microprocessors. IEEE Trans Nucl Sci 49:3100–3106

Sexton FW, Fleetwood DM, Shaneyfelt MR, Dodd PE, Hash GL (1997) Single event gate rupture in thin gate oxides. IEEE Trans Nucl Sci 44:2345–2352

Shannon CE (1948) A mathematical theory of communication. Bell Syst Tech J 27:379–423, Pp 623–656

Shedletsky JJ (1978) Error correction by alternate-data retry. IEEE Trans Comput C-27:106–112

Sherman L (2003) Stratus continuous processing technology – the smarter approach to uptime. Stratus Technologies Whitepaper. Technical report, Stratus Technologies

Shirvani PP, McCluskey EJ (1998) Fault-tolerant systems in a space environment: The CRC ARGOS project. Stanford University, Stanford, CA

Shivakumar P, Kistler M, Keckler SW, Burger D, Alvisi L (2002) Modeling the effect of technology trends on the soft error rate of combinational logic. In: Proceedings of the international conference on dependable systems and networks (DSN 2002), pp. 389–398

Silberberg R, Tsao CH, Letaw JR (1984) Neutron generated single-event upsets in the atmosphere. IEEE Trans Nucl Sci 31:1183–1185

Silvestri M, Gerardin S, Paccagnella A, Ghidini G (2009) Gate rupture in ultra-thin gate oxides irradiated with heavy ions. IEEE Trans Nucl Sci 56:1964–1970

Sklaroff JR (1976) Redundancy management technique for space shuttle computers. IBM J Res Dev 20:20–28

Slayman CW (2005) Cache and memory error detection, correction, and reduction techniques for terrestrial servers and workstations. IEEE Trans Device Mater Reliab 5:397–404

Slegel TJ, Averill RMI, Check MA, Giamei BC, Krumm BW, Krygowski CA, Li WH, Liptay JS, MacDougall JD, McPherson TJ, Navarro JA, Schwarz EM, Shum K, Webb CF (1999) IBM's S/390 G5 microprocessor design. IEEE Micro 19:12–23

Slotnick DL, et al. (1962) The SOLOMON computer, AFIPS '62 (Fall) proceedings, 4–6 Dec 1962, Philadelphia, ACM, New York, pp 97–107

Smith GL (1985) Models for delay faults based on paths. Proc. int. test conf, pp 342–349

Smith M (1997) Application-specific integrated circuits, 1st edn. Addison-Wesley Professional, Boston, MA

Smithsonian Institute (1990) Comp. history. http://americanhistory.si.edu/collections/comphist/

Sogomonian E, Schagaev I (1988) Hardware and software fault tolerance of computer systems. Avtomatika i Telemekhanika 3–39

Squire JS, Palais SM (1963) Physical and logical design of a highly parallel computer. Proc SJCC

Sparc International (1998) The SPARC architecture manual. http://www.gaisler.com/doc/sparcv8.pdf

Srour JR, Marshall CJ, Marshall PW (2003) Review of displacement damage effects in silicon devices. IEEE Trans Nucl Sci 50:653–670

Stassinopoulos EG, Brucker GJ, Calvel P, Baiget A, Peyrotte C, Gaillard R (1992) Charge generation by heavy ions in power MOSFETs, burnout space predictions and dynamic SEB sensitivity. IEEE Trans Nucl Sci 39:1704–1711

Stepanyants A (2001) Fault tolerant processor and its reliability analysis. In: IEEE DSN01, Goteborg

Storey DN (1996) Safety critical computer systems, 1st edn. Prentice Hall, Upper Saddle River, NJ

Stroud CE (2002) A designer's guide to built-in self-test. Springer, Boston, MA

Swift G, Katz R (1996) An experimental survey of heavy ion induced dielectric rupture in Actel Field Programmable Gate Arrays (FPGAs). IEEE Trans Nucl Sci 43:967–972

Swift DW, John GH (1997) Evaluation of the space environment on TOPEX-Poseidon and on-board failure of optocouplers. Document JPL D-14157

Swift GM, Guertin SM (2000) In-flight observations of multiple-bit upset in DRAMs. IEEE Trans Nucl Sci 47:2386–2391

Taber AH, Normand E (1992) Investigation and characterization of SEU effects and hardening strategies in avionics. IBM report 92–L75–020–2. Defense Technical Information Center

Taber A, Normand E (1993) Single event upset in avionics. IEEE Trans Nucl Sci 40:120–126

Takeda E, Kume H, Nakagome Y, Makino T, Shimizu A, Asai S (1983) An As-P double diffused drain MOSFET for VLSI's. IEEE Trans Electron Dev 30:652–657

Tehranipoor M, Peng K, Chakrabarty K (2012) Delay test and small-delay defects, Test and diagnosis for small-delay defects. Springer, New York, NY, pp 21–36

Test method for beam accelerated soft error rate, 2007.

Thambidurai P, Park Y (1988) Interactive consistency with multiple failure modes. In: Proceedings of the seventh symposium on reliable distributed systems. pp 93–100

Turski WM, Wasserman AI (1978) Computer programming methodology. SIGSOFT Softw Eng Notes 3:20–21

Underwood CI (1998) The single-event-effect behaviour of commercial-off-the-shelf memory devices-A decade in low-Earth orbit. IEEE Trans Nucl Sci 45:1450–1457

Valstar JE (1965) The contribution of testability to the cost-effectiveness of a weapon system. IEEE Trans Aerospace AS-3:52–59

Velazco R, Fouillat P, Reis RA, da L (2007) Radiation effects on embedded systems. Springer, New York, NY

Von Neumann J (1956) Probabilistic logics and the synthesis of reliable organisms from unreliable components, Automata studies. Princeton University Press, Princeton, NJ, pp 43–98

Vykhovanets VS (2006) Algebraic decomposition of discrete function Automation and Remote Control March 2006, Volume 67, Issue 3, pp 361–392

Wallmark JT, Marcus SM (1962) Minimum size and maximum packing density of nonredundant semiconductor devices. Proc IRE 50:286–298

Warren KM, Weller RA, Mendenhall MH, Reed RA, Ball DR, Howe CL, Olson BD, Alles ML, Massengill LW, Schrimpf RD, Haddad NF, Doyle SE, McMorrow D, Melinger JS, Lotshaw WT (2005) The contribution of nuclear reactions to heavy ion single event upset cross-section measurements in a high-density SEU hardened SRAM. IEEE Trans Nucl Sci 52:2125–2131

Waskiewicz AE, Groninger JW, Strahan VH, Long DM (1986) Burnout of power MOS transistors with heavy ions of Californium-252. IEEE Trans Nucl Sci 33:1710–1713. doi:10.1109/TNS.1986.4334670

Weaver HT, Axness CL, McBrayer JD, Browning JS, Fu JS, Ochoa A, Koga R (1987) An SEU tolerant memory cell derived from fundamental studies of SEU mechanisms in SRAM. IEEE Trans Nucl Sci 34:1281–1286

Weaver C, Emer J, Mukherjee SS, Reinhardt SK (2004) Techniques to reduce the soft error rate of a high-performance microprocessor. In: Proceedings of the 31st annual international symposium on computer architecture, pp 264–275

Williams HC (2003) The writing on the wall. University of Pennsylvania Press, Philadelphea, PA. ISBN 0-8122-3711-0

Williams TW, Kapur R, Mercer MR, Dennard RH, Maly W (1996) Iddq testing for high performance CMOS-the next ten years. In: European design and test conference proceedings (EDTC 96), pp 578–583

Wirth N (1983) Programming in Modula-2. Springer, Berlin

Wirth N (1988) The programming language Oberon. Softw Pract Exper 18:671–690

Wirth N (1992) Project Oberon: the design of an operating system and compiler. Addison-Wesley, Boston, MA

Wirth GI, Vieira MG, Neto EH, Kastensmidt FL (2008) Modeling the sensitivity of CMOS circuits to radiation induced single event transients. Microelectron Reliab 48:29–36

Wood A (1999) Data integrity concepts, features, and technology. White paper, Tandem Division, Compaq Computer Corporation

Woodard SE, Metze G (1978) Self-checking alternating logic: sequential circuit design, Proceedings of the 5th annual symposium on computer architecture, ISCA'78. ACM, New York, NY, pp 114–122

Wrobel F, Hubert G, Iacconi P (2006) A semi-empirical approach for heavy ion SEU cross section calculations. IEEE Trans Nucl Sci 53:3271–3276

Wyatt RC, McNulty PJ, Toumbas P, Rothwell PL, Filz RC (1979) Soft errors induced by energetic protons. IEEE Trans Nucl Sci 26:4905–4910

Xiaoqing W, Saluja KK, Kinoshita K, Tamamoto H (1996) Equivalence fault collapsing for transistor leakage faults. In: IEEE international workshop on IDDQ testing, pp 79–83

Yen YC (1996) Triple-triple redundant 777 primary flight computer Boeing commercial airplanes. In: 1996 IEEE aerospace applications conference proceedings, vol 1, Seattle, WA. doi:10.1109/AERO.1996.495891

Yu Qingkui, Tang Min, Zhu Hengjing, Zhang Haiming, Zhang Yanwei, Sun Jixing, 2005. Experimental investigation of radiation damage on CCD with protons and cobalt 60 gamma rays. In: 8th European conference on radiation and its effects on components and systems (RADECS 2005), pp LNW2–1–LNW2–5

Zhu X, Baumann R, Pilch C, Zhou J, Jones J, Cirba C (2005) Comparison of product failure rate to the component soft error rates in a multi-core digital signal processor. In: Proceedings of the IEEE international 43rd annualreliability physics symposium, pp 209–214

Ziegler JF (1996) Terrestrial cosmic rays. IBM J Res Dev. doi:10.1147/rd.401.0019

Ziegler JF, Lanford WA (1979) Effect of cosmic rays on computer memories. Science 206:776–788

Ziegler JF, Puchner H (2004) SER—history, trends and challenges: a guide for designing with memory ICs. Cypress

Ziegler JF, Ziegler MD, Biersack JP (2010) SRIM – The stopping and range of ions in matter. Nucl Instr Meth Phys Res 268:1818–1823

Index

V. Castano, I. Schagaev, *Resilient Computer System Design*,
DOI 10.1007/978-3-319-15069-7

FSC
www.fsc.org
MIX
Papier | Fördert
gute Waldnutzung
FSC® C083411

Zeitfracht Medien GmbH
Ferdinand-Jühlke-Straße 7
99095 Erfurt, Deutschland
produktsicherheit@kolibri360.de